Data Science for Neuroimaging

Data Science for Neuroimaging

AN INTRODUCTION

ARIEL ROKEM AND
TAL YARKONI

PRINCETON UNIVERSITY PRESS
PRINCETON AND OXFORD

Published by Princeton University Press
41 William Street, Princeton, New Jersey 08540
99 Banbury Road, Oxford OX2 6JX

press.princeton.edu

All Rights Reserved

Library of Congress Cataloging-in-Publication Data

Names: Rokem, Ariel, 1977– author. | Yarkoni, Tal, author.
Title: Data science for neuroimaging : an introduction / Ariel Rokem and Tal Yarkoni.
Description: Princeton : Princeton University Press, [2024] | Includes bibliographical references and index.
Identifiers: LCCN 2023018370 (print) | LCCN 2023018371 (ebook) | ISBN 9780691222752 (paperback) | ISBN 9780691222738 (hardback) | ISBN 9780691222745 (ebook)
Subjects: MESH: Data Science—methods | Neuroimaging | Programming Languages | Datasets as Topic | BISAC: SCIENCE / Life Sciences / Neuroscience | COMPUTERS / Data Science / Data Analytics
Classification: LCC RC349.D52 (print) | LCC RC349.D52 (ebook) | NLM WL 26.5 | DDC 616.8/04754—dc23/eng/20230801
LC record available at https://lccn.loc.gov/2023018370
LC ebook record available at https://lccn.loc.gov/2023018371

British Library Cataloging-in-Publication Data is available

Editorial: Hallie Stebbins, Kiran Pandey
Jacket/Cover: Wanda España
Production: Lauren Reese
Publicity: William Pagdatoon

Jacket/Cover Credit: Generated using the Parcellation Fragmenter software, created by Kristian Eschenburg, Michael Notter, Claudio Toro-Serey, Amanda Sidwell, Omer Faruk Gulban, and Peer Heerholz

This book has been composed in Arno and sans

10 9 8 7 6 5 4 3 2 1

To my parents, Galit and Freddie   —Ariel

To all lifelong students   —Tal

CONTENTS

PREFACE

This is an exciting time to be working at the intersection of neuroimaging and data science. On the one hand, we have more and more neuroimaging data. Even if you do not have access to a brain scanner yourself, you can gain access to a variety of data sets online and, it seems, start working on your own brain data analysis almost immediately. Some people believe that the mountains of data that we can all now access will yield the kinds of insights that will relieve the pain and suffering due to mental health disorders and neurological diseases. Some even suggest that a better understanding of the biology of differences between individuals will help address the societal inequities that give rise to these differences. These are big hopes, and we carefully share them. On the other hand, as anyone who has started wading into neuroimaging data knows, the reality is often much more mundane. It entails wrangling data between obscure data formats; making different pieces of software work in harmony with each other; it means struggling to make sure that the analysis that you did today still replicates when a collaborator runs it tomorrow, or even when you run it next week; and it means trying to stay up-to-date with the flurry of methods and new findings that are published almost every day.

How do we start to wrap our heads around these challenges? We wrote this book in large part because we wish a book like this had existed when we were coming to terms with these challenges ourselves, during our development as scientists. Fortunately for us, we had mentors and collaborators who could teach us many of the things that are now in this book, but the process of learning was often more roundabout than we wish it had been. As we have started teaching some of these things to others, we once again wished that we had all these pieces of information in one place, and here we are. We sincerely hope that this book will serve to make the path through some of these challenges more direct, or at least a bit more well-illuminated.

We are grateful for the inspiration and help that we received in putting this book together. Some of the material in this book evolved from materials that we put together as part of organizing and teaching the NeuroHackademy Summer Institute in Neuroimaging and Data Science.[1] The book and its contents are inspired by the participants and instructors that have come through the school over the years we have run it, in person (2016–2019, and 2022) and online (2020–2021). We learned a great deal about neuroimaging and data science from them all. The cover image of the book includes

---

1. https://neurohackademy.org

elements generated using the Parcellation Fragmenter software (https://github.com /miykael/parcellation_fragmenter) created by Michael Notter, Kristian Eschenburg, Claudio Toro-Serey, Amanda Sidwell, Peer Herholz, Omer Faruk Gulban, and Ross Markello, when they participated in NeuroHackademy in 2018. We are grateful that they have allowed us to use these beautiful images. We are grateful for the support that Neuro-Hackademy has received from the U.S. National Institute of Mental Health. We are also immensely grateful for the financial, physical, and *intellectual* infrastructure provided by the University of Washington eScience Institute and the wonderful team there. We would also like to thank Teresa Gomez, Sam Johnson, and McKenzie Hagen, who read early versions of the book and made various comments and suggestions, as well as four reviewers for many suggestions on how to improve the content of our book and make it more helpful. To the extent that there are still errors or the book falls short, that is our failing.

We are grateful to the developers of the many excellent open-source software tools that are featured in this book, as well as the communities and institutions that support the development and maintenance of these tools. This book was written almost entirely using Jupyter and the Jupyterbook toolchain. We would like to particularly thank the developers of these tools for enabling new ways of computational expression.

1

Introduction

1.1 Why Data Science?

When you picked up this book to start reading, maybe you were hoping that we would once and for all answer the perennial question: What is data science? Allow us to disappoint you. Instead of providing a short and punchy definition, we are going to try to answer a different question, which we think might be even more important: Why data science? Rather than drawing a clear boundary for you around the topic of data science, this question lets us talk about the reasons that we think that data science has become important for neuroimaging researchers and researchers in other fields, and also talk about the effects that data science has on a broader understanding of the world and even on social issues.

One of the reasons that data science has become so important in neuroimaging is that the *amount* of data that you can now collect has grown substantially in the last few decades. Jack Van Horn, a pioneer of work at the intersection of neuroimaging and data science, described this growth in a paper he wrote a few years ago [Van Horn 2016]. From the perspective of a few years later, he describes the awe and excitement in his lab when, in the early 1990s, they got their first 4 gigabyte (GB) hard drive. A *byte* of data can hold 8 *bits* of information. In our computers, we usually represent numerical data (like the numbers in magnetic resonance imaging [MRI] measurements) using anything from 1 byte (when we are not too worried about the precision of the number) to 8 bytes, or 64 bits (when we need very high precision, a more typical case). The prefix giga-denotes a billion, so the hard drive that Van Horn and his colleagues got could store 4 billion bytes or approximately five hundred million 64-bit numbers. Back in the early 1990s, when this story took place, this was considered a tremendous amount of data, which could be used to store many sessions of MRI data. Today, given the advances that have been made in MRI measurement technology, and the corresponding advances in computing, this would probably store one or maybe two sessions of a high-resolution functional MRI (fMRI) or diffusion MRI (dMRI) experiment (or about half an hour of ultra high-definition video). These advances come hand-in-hand with our understanding that we need more data to answer the kinds of questions that we would like to ask about the brain. There are different ways that this affects the science that we do. If we are interested in answering questions about the brain differences that explain cognitive or behavioral differences between individuals—for example, where in the neuroanatomy do individual differences in structure correspond to the propensity to develop mental health disorders—we are going to need measurements from many individuals to provide sufficient statistical power. Conversely, if we are instead interested in understanding how one brain processes stimuli that come from a very large set—for example, how a particular person's brain represents visual stimuli—we will need to make many measurements in that one brain with as many stimuli as practically possible from that set. These two examples (and there are many others, of course) demonstrate some of the reasons that data sets in neuroimaging are growing. Of course, neuroimaging is not unique in this respect, and additional examples of large data sets in other research domains are given subsequently.

Data Science Across Domains

This book focuses on neuroimaging, but large, complex, and impactful data sets are not unique to neuroscience. Data sets have been growing in many other research fields, and arguably in almost any research field where data is being collected. Here, we will present just a few examples, with a focus on data sets that contain images, and therefore use analysis tools and approaches that intersect with neuroimaging data science in significant ways.

One example of large data sets comes from *astronomy*. Though it has its roots in observations done with the naked eye, in the last few years, measurement methods and data have expanded significantly. For example, the Vera Rubin Observatory,[1] is one project centered on a telescope that has been built at an altitude of about 2,700 m above sea level on the Cerro Pachón ridge in Chile, and is expected to begin collecting data in early 2024. The experiment conducted by the Vera Rubin telescope will be remarkably simple: scanning about half of the night sky every few days and recording it with the highest-resolution digital camera every constructed. However, despite this simplicity, the data that will be produced is unique, both in its scale—the telescope will produce about 20 terabytes (TB) of data *every night*—and its impact. It is anticipated to help answer a range of scientific questions about the structure and nature of the universe. The data, which will eventually reach a volume of approximately 500 petabytes (PB), will be distributed through a specialized database to researchers all over the world, and will serve as the basis for a wide range of research activity in the decades to come. The project is therefore investing significant efforts to build out specialized data science infrastructure, including software and data catalogs.

Similarly, *earth science* has a long tradition of data collection and dissemination. For example, the NASA Landsat program, which focuses on remote sensing the surface of the earth from earth-orbiting satellites has existed since the 1970s. However, recent Landsat missions (the most recent is Landsat 8) have added multiple new types of

1. https://www.lsst.org/

measurements in addition to the standard remote sensing images produced previously. In total, Landsat and related projects produce more than 4 PB of image data every year. Similar to the efforts that are being made to harness these data in astronomy, there are significant attempts to create a data science ecosystem for analysis of these datasets in earth science. One interesting approach is taken by a project called Pangeo,[2] which has created a community platform for big data geoscience, collecting resources, software, and best practices and disseminating them to the earth science community.

Finally, neuroscience is a part of a revolution that is happening across the *biological sciences*. There are many different sources of information that are creating large and complex data sets in biology and medicine, including very large genomic data. However, one particularly potent data-generation mechanism involves new methods for imaging of tissue at higher and higher spatial and temporal resolution, and with more and more coverage. For example, high-throughput methods now can image an entire brain at the resolution of individual synapses. To share data sets related specifically to neuroscience, the US-based Brain Research through Advancing Innovative Neurotechnologies (BRAIN) Initiative has established several publicly-accessible archives. For example, the Distributed Archives for Neurophysiology Data Integration (DANDI) archive[3] shares a range of neurophysiology data, including electrophysiology, optophysiology (measurements of brain activity using optical methods), and images from immunostaining experiments. At the time of writing, the archive has already collected 440 TB of data of this sort, shared by researchers all over the US and the world.

2. https://pangeo.io/
3. https://dandiarchive.org/

In addition to the sheer volume of data, the *dimensionality* of the data is also increasing. This is in part because the kinds of measurements that we can make are changing with improvements in measurement technologies. This is related to the volume increase that we mentioned previously but is not the same. It is one thing to consider where you will store a large amount of data and how you might move it from one place to the other. It is a little bit different to understand data that are now collected at a very high resolution, possibly with multiple different complementary measurements at every time point, at every location, or for every individual. This is best captured in the well-known term "the curse of dimensionality," which describes how data of high dimensionality defy our intuitions and expectations based on our experiences with low-dimensional data. As with the volume of the data, the issue of dimensionality is common to many research fields and tools to understand high-dimensional data have been developed in many of these fields.

This brings us to another reason that data science is important. This is because we can gain a lot from borrowing methods from other fields in which data has become ubiquitous, or from fields that are primarily interested in data, such as statistics and some parts of computer science and engineering. These fields have developed many interesting methods for dealing with large and high-dimensional data sets, and these *interdisciplinary* exchanges have proven very powerful. Researchers in neuroscience have been very successful in applying relatively new techniques from these other fields to neuroscience data. For example, *machine learning* methods have become quite popular in analyzing high-dimensional data sets and have provided important insights about a variety of research questions. Interestingly, as we will see in some of the chapters ahead, the exchange has not been completely one-sided, and neuroscientists have also been able to contribute to the conversation about data analysis in interesting and productive ways.

Another way in which data science has contributed to improvements in neuroscience is through an emphasis on *reproducibility*. Reproducibility of research findings requires ways to describe and track the different phases of research in a manner that would allow others to precisely repeat it. Thus the data needs to be freely available, and the code used to analyze the data needs to be available in such a way that others can also run it. This is facilitated by the fact that many of the important tools that are central to data science are *open-source* software tools that can be inspected and used by anyone. This means that other researchers can scrutinize the results of the research from top to bottom, and understand them better. It also means that the research can be more easily extended by others, increasing its impact.

Data science matters because once we start dealing with large and complex data sets, especially if they are collected from human subjects, the *ethical* considerations for use of the data and its potential harm to individuals and communities changes quite a bit. For example, considerations of potential harms need to go beyond just issues related to privacy and the protection of private information. Privacy is of course important, but some of the harms of large-scale data analysis may befall individuals who are not even in the data. For example, individuals who share certain traits or characteristics of the individuals who are included in the data could be affected, as could those individuals who were not included in the data. How large biomedical data sets are being collected and the implications of the decisions made in designing these studies have profound implications for how the conclusions apply to individuals across society.

Taken together, these factors make data science an important and central part of contemporary scientific research. However, learning about data science can be challenging, even daunting. How can neuroimaging researchers productively engage with these topics? This brings us to our next subject: the intended audience for this book.

1.2 Who This Book Is For

This book was written to introduce researchers and students in a variety of research fields to the intersection of data science and neuroimaging. Neuroimaging gives us a view into the structure and function of the living human brain, and it has gained a solid foothold in research on many different topics. As a consequence, people who use neuroimaging to study the brain come to it from many different backgrounds and with many different questions in mind. We wrote this book thinking of a variety of situations in which additional technical knowledge and fluency in the language of data science can provide a benefit to individuals—whether it be in rounding out their training, in enabling a research direction that would not be possible otherwise, or in facilitating a transition in their career trajectory. The researchers we have written this book for usually have some background in neuroscience and this book is not meant to provide an introduction to neuroscience. When we refer to specific neuroscience concepts and measurements we might explain them, but for a more comprehensive introduction to neuroscience and neuroimaging, we recommend picking up another book (of which there are many); all chapters, including this one, end with an "Additional Resources" section that will include pointers to these resources. We will also not discuss how neuroimaging data comes about. Several excellent textbooks describe the physics of signal formation in different neuroimaging modalities and considerations in neuroimaging data collection and experimental design. Finally, we will present certain approaches to the analysis of neuroimaging data, but this is also not a book about the statistical analysis of neuroimaging data. Again, we refer readers to another book specifically on this topic. Instead, this book aims to give a broad range of researchers an initial entry point to data science tools and approaches and their application to neuroimaging data.

1.3   How We Wrote This Book

This book reflects our own experience of doing research at the intersection of data science and neuroimaging and it is based on our experience working with students and collaborators who come from a variety of backgrounds and have different reasons for wanting to use data science approaches in their work. The tools and ideas that we chose to write about are all tools and ideas that we have used in some way in our research. Many of them are tools that we use daily in our work. This was important to us for a few reasons: the first is that we want to teach people things that we find useful. Second, it allowed us to write the book with a focus on solving specific analysis tasks. For example, in many of the chapters, you will see that we walk you through ideas while implementing them in code, and with data. We believe that this is a good way to learn about data analysis because it provides a connecting thread from scientific questions through the data and their representation to generating specific answers to these questions. Finally, we find these ideas compelling and fruitful. That is why we were drawn to them in the first place. We hope that our enthusiasm for the ideas and tools described in this book will be infectious enough to convince the readers of their value.

1.4   How You Might Read This Book

More important than how we wrote this book, however, is how we envision you might read it. The book is divided into several parts.

Data science operates best when the researcher has comprehensive, explicit, and fine-grained control. The first part of the book introduces some fundamental tools that give users such control. These serve as a base layer for interacting with the computer and are generally applicable to whatever data analysis task we might perform. Operating with tools that give you this level of control should make data analysis more pleasant and productive, but it does come with a bit of a learning curve that you will need to climb. This part will hopefully get you up part of the way, and starting to use these tools in practice should help you to get up the rest of the way. We will begin with the Unix operating system and the Unix *command line interface* (in Chapter 2). This is a computing tool with a long history, but it is still very well suited for flexible interaction with the computer's operating system and file system, as its robustness and efficiency have been established and honed over decades of application to computationally intensive problems in scientific computing and engineering. We will then (in chapter 3) introduce the idea of *version control*—a way to track the history of a computational project—with a focus on the widely used git version control system. Formal version control is a fundamental building block of data science as it provides fine-grained and explicit control over the versions of software that a researcher works with, and also facilitates and eases collaborative work on data analysis programs. Similarly, computational *environments* and computational *containers*, introduced in chapter 4, allow users to specify the different software components that they use for a specific analysis while preventing undesirable interactions with other software.

The book introduces a range of tools and ideas, but within the broad set of ways to engage with data science, we put a particularly strong emphasis on programming. We think that programming is an important part of data science because it is a good way to apply quantitative ideas to large amounts of data. One of the major benefits of programming over other approaches to data analysis, such as applications that load data and allow you to perform specific analysis tasks at the click of a button, is that you are given the freedom to draw outside the lines: with some effort, you can implement any quantitative idea that you might come up with. Conversely, when you write a program to analyze your data, you have to write down exactly what happens with the data and in what order. This supports the goal of automation; you can run the same analysis on multiple data sets. It also supports the goal of making the research reproducible and extensible. That is because it allows others to see what you did and repeat it in exactly the same way. That means programming is central to many of the topics we will cover. The examples that are provided will use neuroimaging data, but as you will see, many of these examples could have used other data just as well. Note that this book is not meant to be a general introduction to programming. We are going to spend some time introducing the reader to programming in the *Python programming language* (starting in chapter 5; we will also explain specifically why we chose the Python programming language for this book), but for a gentler introduction to programming, we will refer you to other resources. However, we will devote some time to things that are not usually mentioned in books about programming but are crucially important to data science work, such as how to test software and profile its performance, and how to effectively share software with others.

In the next two parts of the book, we will gradually turn towards topics that are more specific to data science in the context of scientific research, and neuroimaging in particular.

First, we will introduce some general-purpose scientific computing tools for numerical computing (in chapter 8), data management and exploration (chapter 9), and data visualization (chapter 10). Again, these tools are not neuroimaging-specific, but we will focus in particular on the kinds of tasks that will be useful when working with neuroimaging data. Then, in the next part, we will describe in some detail tools that are specifically implemented for work with neuroimaging data: the breadth of applications of Python to neuroimaging data will be introduced in chapter 11. We will go into further depth with the NiBabel software library, which gives users the ability to read, write, and manipulate data from standard neuroimaging file formats in chapter 12 and chapter 13.

The last two parts of the book explore in more depth two central applications of data science to neuroimaging data. In the first of these (Part V), we will look at image processing, introducing general tools and ideas for understanding image data (in chapter 14), and focusing on tasks that are particularly pertinent for neuroimaging data analysis; namely, image segmentation (in chapter 15) and image registration (in chapter 16). Finally, the last part of the book will provide an introduction to the broad field of machine learning (Part VI). Both of these applications are taken from fields that could fill entire textbooks. We have chosen to provide a path through these that emphasizes an intuitive understanding of the main concepts, with code used as a means to explain and explore these concepts.

Throughout the book, we provide detailed examples that are spelled out in code, with relevant data sets. This is important because the ideas we will present can seem arcane or obscure if only their mathematical definitions are provided. We feel that a software implementation that lays out the steps that are taken in analysis can help demystify them and provide clarity. If the description or the (rare) math that describes a particular idea is opaque, we hope that reading through the code that implements the idea will help readers understand it better. We recommend that you not only read through the code but also run the code yourself. Even more important to your understanding, try changing the code in various ways and rerunning it with these changes. We propose some variations in the sections labeled "Exercise" that are interspersed throughout the text. Solutions to these exercises are provided in Appendix 1. Some code examples are abbreviated by calling out to functions that we implemented in a companion software library that we named `ndslib`. This library includes functions that download and make relevant data sets available within the book's chapters. We provide a reference to functions that are used in the book in Appendix 2 and we also encourage the curious reader to inspect the code that is openly available on GitHub.[4]

1.4.1 Jupyter

One of the tools that we used to write this book, and that we hope that you will use in reading and working through this book, is called Jupyter.[5] The Jupyter notebook is an

---

4. https://github.com/neuroimaging-data-science/ndslib
5. https://jupyter.org/

application that weaves together text, software, and results from computations. It is very popular in data science and widely used in scientific research. The notebook provides fields to enter text or code—these are called *cells*. Code that is written in a code cell can be sent to an interactive programming language interpreter for evaluation. This interpreter is referred to as the *kernel* of this notebook. For example, the kernel of the notebook can be an interactive Python session. When you write a code cell and send it to the kernel for evaluation, the Python interpreter runs the code and stores the results of the computation in its memory for as long as the notebook session is maintained (so long as the kernel is not restarted). That means that you can view these results and also use these results in the following code cells. The creators of the Jupyter notebook, Brian Granger and Fernando Pérez, recently explained the power of this approach in a paper that they wrote [Granger and Pérez 2021]. They emphasize something that we hope that you will learn to appreciate as you work through the examples in this book, which is that data analysis is a collaboration between a person and their computational environment. Like other collaborations, it requires a healthy dialogue between both sides. One way to foster this dialogue is to work in an environment that makes it easy for the person to perform a variety of different tasks: analyze data, of course, but also explore the data, formulate hypotheses and test them, and also play. As they emphasize, because the interaction with the computer is done by writing code, in the Jupyter environment a single person is both the author and the user of the program. However, because the interactive session is recorded in the notebook format together with rich visualizations of the data (you will learn how to visualize data in chapter 10) and interactive elements, the notebooks can also be used to communicate their findings with collaborators or publish these results as a document or a webpage. Indeed, most of the chapters of this book were written as Jupyter notebooks that weave together explanations with code and visualizations. This is why you will see sections of code, together with the results of running that code interwoven with explanatory text. This also means that you can repeat these calculations on your computer and start altering, exploring, and playing with them. All of the notebooks that constitute the various chapters of this book can also be downloaded from the book website.[6]

1.4.2   Setting Up

To start using Jupyter and to run the contents of the notebooks that constitute this book, you will need to set up your computer with the software that runs Jupyter and also with the software libraries that we use in the different parts of the book. Setting up your computer will be much easier after you gain some familiarity with the set of tools introduced in the next part of the book. For that reason, we put the instructions for setup and for running the code in Section 4.3, at the end of the chapter that introduces these tools. If you are keen to get started, read through the next chapter and you will eventually reach these instructions.

---

6. http://neuroimaging-data-science.org

1.5   Additional Resources

For more about the fundamentals of MRI, you can refer to one of the following:

> D McRobbie, E Moore, M Graves, and M Prince. *MRI from Picture to Proton* (3rd ed.). Cambridge University Press, 2017.
>
> S A Huettel, A W Song, and G McCarthy. *Functional Magnetic Resonance Imaging*. Sinauer Associates, 2014.

For more about the statistical analysis of MRI data, we refer the readers to the following:

> R A Poldrack, J A Mumford, and T E Nichols. *Handbook of Functional MRI Analysis*. Cambridge University Press, 2011.

This book will touch on data science ethics in only a cursory way. This topic deserves further reading and there are fortunately several great books to read. We recommend the following two:

> C D'Ignazio and L Klein. *Data Feminism*. MIT Press, 2020.
>
> C O'Neil. *Weapons of Math Destruction: How Big Data Increases Inequality and Threatens Democracy*. Broadway Books, 2016.

For more about the need for large data sets in neuroimaging, you can read some of the papers that explore the statistical power of studies that examine individual differences [Button et al. 2013]. In a complementary opinion, Thomas Naselaris and his colleagues demonstrate how sometimes we don't need many subjects, but instead would rather opt for a lot of data on each individual [Naselaris et al. 2021].

PART I
The Data Science Toolbox

2

The Unix Operating System

The Unix operating system is a fundamental tool for controlling the way that the computer executes programs, including in data analysis. Its history goes back to the 1960s, but it is still one of the most commonly used tools for computing today. In the 1960s, when Unix was first developed, computers were very different, accessed usually as mainframe systems that were shared among multiple users. Each user would connect to the computer through their terminal device: a keyboard and a screen that were used to send instructions to the computer and receive outputs. This was before the invention of the first graphical user interfaces (GUIs), so all of the instructions had to be sent to the computer in the form of text. At that time, the main way to interact with the computer was through an application called a *shell*. The shell application usually includes a prompt, where users can type a variety of different commands. This prompt is also called a *command line*. The commands typed at the command line can be sent to the computer's operating system to do a variety of different operations, or to launch various other programs. Often, these programs will then produce text outputs that are printed into the shell as well. Over the years, computers evolved and changed, but the idea of a text-based terminal remains. In different operating systems, you will find the shell application in different places. On Windows, you can install a shell application by installing git for Windows[1] (this will also end up being useful for the following sections, which introduce version control with git and containerization with Docker). On Apple's Mac computers, you can find a shell application in your Applications/Utilities folder under the name "Terminal." On the many variants of the Linux operating system, the shell is quite central as well and comes installed with the operating system.

2.1 Using Unix

The developers of the Unix shell believed that programs that run in this kind of environment should be built to each do only one thing. Ideally, each program's output should be formatted so that it could be used as input to another program. This means that users could use multiple small programs to construct more complicated programs and pipelines based

---

1. https://gitforwindows.org/

13

on combinations of different tools. Let's look at a simple example. Opening up the shell, you will be staring at the prompt. On my computer that looks something like this:

```
$
```

You can type commands into this prompt and press the "enter" or "return" key to execute them. The shell will send these commands to the operating system and then some output might appear inside the shell. This is what is called a read-evaluate-print loop, or REPL. That is because the application reads what you type; and evaluates it to understand what it means and what information to provide in return; prints that information to the screen; and then it repeats that whole process, in an infinite loop, which ends only when you quit the application or turn off the computer.

2.1.1   Exploring the File System

When you first start the shell, the working directory that it immediately sees is your home directory. This means that the files and folders in your home directory are immediately accessible to you. For example, you can type the `ls` command to get a listing of the files and folders that the shell sees in your working directory. For example, in the shell on one of our computers:

```
$ ls
Applications    Downloads    Music       Untitled.ipynb tmp
Desktop         Library      Pictures    miniconda3
Documents       Movies       Public      projects
```

Most of the items listed here are folders that came installed with Ariel's computer when he bought it. One example is `Documents` and `Desktop`, while others are folders that he created in the home directory; for example, `projects`. There is also a single stray file here, `Untitled.ipynb`, which is a lone Jupyter notebook (`ipynb` is the extension for these files) that remains here from some impromptu data analysis that he once did. The `ls` command (and many other Unix commands) can be modified using *flags*. These are characters or words added to the command that modify the way that the command runs. For example, if adding the `-F` flag to the call to `ls`, a slash (`/`) character is added at the end of the names of folders, which is practically useful because it tells us which of the names in the list refer to a file and which refer to a folder that contains other files and folders.

```
$ ls -F
Applications/    Downloads/    Music/       Untitled.ipynb tmp/
Desktop/         Library/      Pictures/    miniconda3/
Documents/       Movies/       Public/      projects/
```

In general, if we want to know more about a particular command, we can issue the `man` command, followed by the name of the command for which we would like to read the so-called `man` page (`man` presumably stands for "manual"). For example, the following command would open the `man` page for the `ls` command, telling us about all of the options that we have to modify the way that `ls` works.

```
$ man ls
```

To exit the man page, we would type the q key. We can ask the shell to change the working directory in which we are currently working by issuing the `cd` (i.e., "change directory") command,

```
$ cd Documents
```

which would then change what it sees when we ask it to list the files.

```
$ ls -F
books/
conferences/
courses/
papers/
```

This is the list of directories that are within the `Documents` directory. We can see where the change has occurred by asking the shell where we are, using the `pwd` command (which stands for "present working directory").

```
$ pwd
/Users/arokem/Documents
```

Note that this is the answer that Ariel sees (on his Mac laptop computer) and you might see something slightly different depending on the way your computer is set up. For example, if you are using the shell that you installed from git for Windows on a Windows machine (this shell is also called a `git bash` shell), your home directory is probably going to look more like this:

```
$ pwd
/c/Users/arokem
```

This is the address of the standard `C:\Users\arokem` Windows home directory, translated into a more Unix-like format. If we want to change our working directory back to where the shell started, we can call `cd` again. This command can be used in one of several ways:

```
$ cd /Users/arokem
```

This is a way to refer to the *absolute path* of the home directory. It is *absolute* because this command would bring us back to the home directory, no matter where in the file system we happened to be working before we issued it. This command also tells us where within the structure of the file system the home directory is located. The slash (/) characters in the command are to be read as separators that designate the relationships between different items within the file system. For example, in this case, the home directory is considered to be inside of a directory called "Users," which in turn is inside the *root* of the file system (simply designated as the / at the beginning of the absolute path). This idea— that files and folders are inside other files and folders—organizes the file system as a whole. Another way to think about this is that the file system on our computer is organized as a tree. The root of the tree is the root of the entire file system (/) and all of the items saved in the file system stem from the root. Different branches split off from the root, and they can split further. Finally, at the end of the branches (the leaves, if you will) are the files that are saved within folders at the end of every path through the branches of the tree. The tree structure makes organizing and finding things easier.

Another command we might issue here that would also take us to the home directory is:

```
$ cd ..
```

The .. is a special way to refer to the directory directly containing the directory in which we are currently working within the file system tree, bringing us one step closer to the root. Depending on what directory you are already in, it would take you to different places. Because of that, we refer to this as a *relative path*. Similarly, if we were working within the home directory and issued the following command

```
$ cd Documents
```

this would be a relative path. This is because it does not describe the relationship between this folder and the root of the file system, and would operate differently depending on our present working directory. For example, if we issue that command while our working directory was inside of the Documents directory, we would get an error from the shell because, given only a relative path, it cannot find anything called Documents inside of the Documents folder.

One more way we can get to the home directory from anywhere in the file system is by using the tilde character (~). So, this command

```
$ cd ~
```

is equivalent to this command.

```
$ cd /Users/arokem
```

Similarly, you can refer to folders inside of your home directory by attaching the ~ /
before writing down the relative path from the home directory to that location. For example, to get from anywhere to the Documents directory, I could issue the following,
which is interpreted as an absolute path.

```
$ cd ~/Documents
```

Exercise

The touch command creates a new empty file in your file system. You can create it using
relative and absolute paths. How would you create a new file called new_file.txt
in your home directory? How would you do that if you were working inside of your
~/Documents directory? The mv command moves a file from one location to another.
How would you move new_file.txt from the home directory to the Documents
directory? How would this be different from using the cp command? (Hint: use the mv
and cp man pages to see what these commands do).

2.1.2 The Pipe Operator

There are many other commands in the Unix shell, and we will not demonstrate all of them
here (see a table below of some commonly used commands). Instead, we will now proceed
to demonstrate one of the important principles that we mentioned before: the output of
one command can be directly used as an input to another command. For example, if we
wanted to quickly count the number of items in the home directory, we could provide the
output of the ls command directly as input into the wc command (i.e., "word count").
In Unix, that is called creating a pipe between the commands and we use the *pipe operator*,
which is the vertical line that usually sits in the top right of the US English keyboard: |, to
do so:

```
$ ls  | wc
    13      13      117
```

Now, instead of printing out the list of files in my home directory, the shell prints out
the number of words, lines, and characters in the *output* of the ls command. It tells us that
there are thirteen words, thirteen lines, and 117 characters in the output of ls (it looked
as though the output of ls was three lines, but there were line breaks between each column in the output). The order of the pipe operation is from left to right, and we do not

have to stop here. For example, we can use another pipe to ask how many words in the output of ls contain the letter D. For this, we will use the grep command, which finds all of the instances of a particular letter or phrase in its input:

```
$ ls | grep "D" | wc
      3       3     28
```

To see why this is the case, try running this yourself in your home directory. Try omitting the final pipe to wc and seeing what the output of ls|grep "D" might be. This should give you a sense of how Unix combines operations. Though it may seem a bit silly at first—why would we want to count how many words have the letter "D" in them in a list of files?—when combined in the right way, it can give you a lot of power to automate operations that you would like to do, e.g., identifying certain files in a directory and processing all of them through a command line application.

Command	Description
ls	List the contents of the current directory.
cd	Change the current directory.
pwd	Print the path of the current directory.
mkdir	Create a new directory.
touch	Create a new file.
cp	Copy a file or directory.
mv	Move or rename a file or directory.
rm	Remove a file or directory.
cat	Print the contents of a file to the terminal.
less	View the contents of a file one page at a time.
grep	Search for a pattern in a file or files.
sort	Sort the lines of a file.
find	Search for files based on their name, size, or other attributes.
wc	Print the number of lines, words, and bytes in a file.
chmod	Change the permissions of a file or directory.
chown	Change the ownership of a file or directory.
head	Print the first few lines of a file.
tail	Print the last few lines of a file.
diff	Compare two files and show the differences between them.

2.2   More About Unix

We are going to move on from Unix now. We provide a table with a few of the Unix commands that you have already seen, as well as a few more, as a starting point for your independent exploration. In the next two chapters, you will see some more intricate and specific uses of the command line through two applications that run as command-line interfaces. Together with and also independently of the tools you will see, Unix provides a lot of power to explicitly operate on files and folders in your file system and to run a variety of applications, so becoming fluent with the shell will be a boon to your work.

2.3   Additional Resources

To learn more about the Unix philosophy, we recommend *The Unix Programming Environment* by Brian Kernighan and Rob Pike. The authors are two of the original developers of the system. The writing is a bit archaic, but understanding some of the constraints that applied to computers at the time that Unix was developed could help you understand and appreciate why Unix operates as it does, and why some of these constraints have been kept, even while computers have evolved and changed in other respects.

3

Version Control

Another set of tools that give us more control over our computational environment are those that track changes in our software and analysis. If this seems like a boring or trivial task to you, consider the changes you need to make to a program that you create to analyze your data throughout a long project. Also, consider what happens when you work with others on the same project, and multiple different people can introduce changes to the code that you are using. If you have ever tried to do either of these things, you may have already invented a way to track changes to your programs. Two examples of this are naming the files with the date of the most recent change or adding the name of the author who made the most recent change to the filename. This is a form of version control, but it is often idiosyncratic and error-prone. Instead, software tools for version control allow us to explicitly add a particular set of changes to the code that we are writing as we go along, keeping track of who made which changes and when. They also allow us to merge changes that others are doing. If needed, it marks where changes we have made conflict with changes that others have made, and allows us to resolve these conflicts before moving on.

3.1   Getting Started with Git

One of the most widely used software for version control is called Git. Without going too much into the difference between Git and other alternatives, we will just mention here that one of the reasons that Git is very widely used is because of the availability of services that allow you to use your version control to share your code with others through the web and to collaborate with them on the same code-base fairly seamlessly. A couple of the most popular services are GitHub[1] and GitLab.[2] After we introduce Git and go over some of the basic mechanics of using Git, we will demonstrate how you can use one of these services to collaborate with others and to share your code openly with anyone. This is an increasingly common practice and one that we will talk about more in chapter 7.

---

1. https://github.com
2. https://gitlab.com

Git is a command-line application. Like the ls, cd, and pwd Unix commands that you saw in chapter 2, we use it in a shell application.

Installing Git

First of all, verify whether you need to install Git. If you already have a Unix terminal to work with, you might already have Git installed. In particular, if you installed Git for windows as a way to get the Unix shell in Section 2, you are already set. Type git at the shell to check. If you get a long help document, this means that you have Git installed. Otherwise, follow the instructions in the Git online documentation[3] to install the right version for your computer's operating system.

3. https://git-scm.com/book/en/v2/Getting-Started-Installing-Git

---

In contrast to the other shell commands you saw before, Git can do many different things. To do that, Git has subcommands. For example, before you start using Git, you need to configure it to recognize who you are using the git config subcommand. We need to tell Git both our name and our email address.

```
$ git config --global user.name "Ariel Rokem"
$ git config --global user.email arokem@gmail.com
```

This configuration step only needs to be done once on every computer on which you use Git, and is stored in your home directory (in a file called ~/.gitconfig).

Exercise

Download and install Git and configure it with your name and email address.

The next subcommand that we will usually use with Git is the one that you will start your work with from now on: the subcommand to initialize a *repository*. A repository is a folder within your file system that will be tracked as one unit. That is, your different projects can be organized into different folders, and each one of them should probably be tracked separately. So one project folder becomes one Git repository (also often referred to as a repo for short). Before starting to use Git, we then make a directory in which our project code will go.

```
$ mkdir my_project
$ cd my_project
```

The Unix mkdir command creates a new directory called (in this case) my_project and the cd command changes our working directory so that we are now working within the new directory that we have created. And here comes the first Git subcommand, to initialize a repository:

```
$ git init
Initialized empty Git repository in /Users/arokem/projects/my_project/.git/
```

As you can see, when this subcommand is issued, Git reports back that the empty repository has been created. We can start adding files to the repository by creating new files and telling Git that these are files that we would like to track.

```
$ touch my_file.txt
$ git add my_file.txt
```

We have already seen the `touch` Unix command before, it creates a new empty file. Here, we have created a text file, but this can also be (more typically) a file that contains a program written in some programming language. The `git add` command tells Git this is a file that we would like to track, and we would like Git to record any changes that we make in this file. A useful command that we will issue early and often is the `git status` command. This command reports to you about the current state of the repository that you are working in, without making any changes to its content or state.

```
$ git status
On branch master

No commits yet

Changes to be committed:
(use "git rm --cached <file>..." to unstage)
   new file:   my_file.txt
```

Here it is telling us a few things about the repository and our work within it. Let's break this message down line by line. First of all, it tells us that we are working on a *branch* called `master`. Branches are a very useful feature of Git. They can be thought of as different versions of the same repository that you are storing side by side. The default when a repository is initialized is to have only one branch called `master`. However, in recent years, Git users have raised objections to the use of this word to designate the default branch, together with objections to other uses of "master/slave" designations in a variety of computer science and engineering (e.g., in distributed computing and databases. To read more about this, see "Additional Resources" at the end of this section). Therefore, we will immediately change the name of the branch to "`main`."

```
$ git branch -M main
```

In general, it is good practice not to use the default branch (now called `main`), but a different branch for all of your new development work. The benefit of this approach is that as you are working on your project, you can always come back to a clean version of your work that is stored on your main branch. It is only when you are sure that you are ready to incorporate some new development work that you *merge* a new development branch into your main branch. The next line in the status message tells us that we have not made any commits to the history of this repository yet. Finally, the last three lines tell us that there are changes that we could commit and make part of the history of the repository. This is the new file that we just added, and is currently stored in something called the *staging area*. This is a part of the repository into which we have put files and changes that we might put into the history of the repository next, but we can also *unstage* them so that they are not included in the history. Either because we do not want to include them yet, or because we do not want to include them at all. In parentheses Git also gives us a hint about how we might remove the file from the staging area, that is, how we could unstage it. But for now, let's assume that we do want to store the addition of this file in the history of the repository and track additional changes. Storing changes in the Git repository is called *making a commit*, and the next thing we do here is to issue the git commit command.

```
$ git commit -m "Adds a new file to the repository"
[main (root-commit) 1808a80] Adds a new file to the repository
1 file changed, 0 insertions(+), 0 deletions(-)
create mode 100644 my_file.txt
```

As you can see, the commit command uses the -m flag. The text in quotes after this flag is the *commit message*. We are recording a history of changes in the files that we are storing in our repository, and alongside the changes, we are storing a log of these messages that give us a sense of who made these changes, when they made them, and (stored in these commit messages) what the intention of these changes was. Git also provides some additional information: a new file was created, and the mode of the file is 664, which is the mode for files that can be read by anyone and written by the user (for more about file modes and permissions, you can refer to the man page of the chmod Unix command). There were zero lines inserted and zero lines deleted because the file is empty and does not have any text in it yet.

We can revisit the history of the repository using the git log command.

```
$ git log
```

This should drop you into a new buffer within the shell that looks something like this:

```
commit 1808a8053803150c8a022b844a3257ae192f413c (HEAD -> main)
Author: Ariel Rokem <arokem@gmail.com>
Date:    Fri Dec 31 11:29:31 2021 -0800

    Adds a new file to the repository
```

It tells you who made the change (Ariel, in this case), how you might reach that person (yes, that is Ariel's email address; Git knows it because of the configuration step we did when we first started using Git) and when the change was made (almost noon on the last day of 2021, as it so happens). In the body of that entry, you can see the message that we entered when we issued the `git commit` command. At the top of the entry, we also have a line that tells us that this is the commit identified through a string of letters and numbers that starts with "1808a." This string is called the SHA of the commit, which stands for "secure hash algorithm". Git uses this clever algorithm to encode all of the information included in the commit (in this case, the name of the new file, and the fact that it is empty) and to encrypt this information into the string of letters and numbers that you see. The advantage of this approach is that this identifier is unique within this repository because this specific change can only be made once.[4] This allows us to look up specific changes that were made within particular commits, and also allows us to restore the repository, or parts of it, back to the state they were in when this commit was made. This is a particularly compelling aspect of version control: you can effectively travel back in time to the day and time that you made a particular change in your project and visit the code as it was on that day and time. This also relates to the one last detail in this log message: next to the SHA identifier, in parentheses is the message "(`HEAD -> main`)." The word "HEAD" always refers to the state of the repository in the commit you are currently viewing. Currently, HEAD—the current state of the repository—is aligned with the state of the `main` branch, which is why it is pointing to `main`. As we continue to work, we will be moving HEAD around to different states of the repository. We will see the mechanics of that in a little bit. But first, let's leave this log buffer by typing the q key.

Let's check what Git thinks of all this.

```
$ git status
On branch main
nothing to commit, working tree clean
```

We are still on the main branch. But now Git tells us that there is nothing to commit and that our working tree is clean. This is a way of saying that nothing has changed since the last time we made a commit. When using Git for your own work, this is probably the state in which you would like to leave it at the end of a working session. To summarize so far: we have not done much yet, but we have initialized a repository and added a file to it. Along the way, you have seen how to check the status of the repository, how to commit changes to the history of the repository, and how to examine the log that records the history of the repository. As you will see next, at the most basic level of operating with Git, we go through cycles of changes and additions to the files in the repository and commits

---

4. It is *theoretically* possible to create another change that would have the same SHA, but figuring out how to do that would be so computationally demanding that, while theoretically possible, it is not something we need to worry about in any practical sense.

that record these changes, both operations that you have already seen. Overall, we will discuss three different levels of intricacy. The first level, which we will start discussing next, is about tracking changes that you make to a repository and using the history of the repository to recover different states of your project. At the second level, we will dive a little bit more into the use of branches to make changes and incorporate them into your project's `main` branch. The third level is about collaborating with others on a project.

3.2   Working with Git at the First Level: Tracking Changes That You Make

Continuing the project we started previously, we might make some edits to the file that we added. The file that we created, `my_file.txt`, is a text file and we can open it for editing in many different applications (we discuss text editors, which are used to edit code, in chapter 6). We can also use the Unix `echo` command to add some text to the file from the command line.

```
$ echo "a first line of text" >> my_file.txt
```

Here, the » operator is used to redirect the output of the `echo` command (you can check what it does by reading the man page) into our text file. Note that there are two > symbols in that command, which indicates that we would like to concatenate the string of characters to the end of the file. If we used only one, we would overwrite the contents of the file entirely. For now, that is not too important because there was not anything in the file, to begin with but it will be important in what follows.

We have made some changes to the file. Let's see what Git says:

```
$ git status
On branch main
Changes not staged for commit:
(use "git add <file>..." to update what will be committed)
(use "git restore <file>..." to discard changes in working directory)
    modified:   my_file.txt

no changes added to commit (use "git add" and/or "git commit -a")
```

Let's break that down again. We are still on the `main` branch. But we now have some changes in the file, which are indicated by the fact that there are changes not staged for a commit in our text file. Git provides two hints, in parentheses, as to different things we might do next, either to move these changes into the staging area or to discard these changes and return to the state of affairs as it was when we made our previous commit.

Let's assume that we would like to retain these changes. That means that we would like to first move these changes into the staging area. This is done using the `git add` subcommand.

```
$ git add my_file.txt
```

Let's see how this changes the message that `git status` produces.

```
$ git status
On branch main
Changes to be committed:
(use "git restore --staged <file>..." to unstage)
    modified:   my_file.txt
```

The file in question has moved from `changes not staged for commit` to `changes to be committed`. That is, it has been moved into the staging area. As suggested in the parentheses, it can be unstaged, or we can continue as we did before and commit these changes.

```
$ git commit -m"Adds text to the file"
[main 42bab79] Adds text to the file
1 file changed, 1 insertion(+)
```

Again, Git provides some information: the branch into which this commit was added (`main`) and the fact that one file changed, with one line of insertion. We can look at the log again to see what has been recorded:

```
$ git log
```

This should look something like the following, with the most recent commit at the top.

```
commit 42bab7959c3d3c0bce9f753abf76e097bab0d4a8 (HEAD -> main)
Author: Ariel Rokem <arokem@gmail.com>
Date:   Fri Dec 31 20:00:05 2021 -0800

    Adds text to the file

commit 1808a8053803150c8a022b844a3257ae192f413c
Author: Ariel Rokem <arokem@gmail.com>
Date:   Fri Dec 31 11:29:31 2021 -0800

    Adds a new file to the repository
```

Again, we can move out of this buffer by pressing the q key. We can also check the status of things again, confirming that there are no more changes to be recorded.

```
$ git status
On branch main
nothing to commit, working tree clean
```

This is the basic cycle of using Git: make changes to the files, use `git add` to add them to the staging area, and then `git commit` to add the staged changes to the history of the project, with a record made in the log. Let's do this one more time, and see a couple more things that you can do along the way, as you are making changes. For example, let's add another line of text to that file.

```
$ echo "another line of text" >> my_file.txt
```

We can continue in the cycle, with a `git add` and then `git commit` here, but let's pause to introduce one more thing that Git can do, which is to tell you exactly what has changed in your repository since your last commit. This is done using the `git diff` command. If you issue that command

```
$ git diff
```

you should be again dropped into a different buffer that looks something like this:

```
diff --git a/my_file.txt b/my_file.txt
index 2de149c..a344dc0 100644
--- a/my_file.txt
+++ b/my_file.txt
@@ -1 +1,2 @@
a first line of text
+another line of text
```

This buffer contains the information that tells you about the changes in the repository. The first few lines of this information tell you a bit about where these changes occurred and the nature of these changes. The last two lines tell you exactly the content of these changes; i.e., the addition of one line of text (indicated by a "+" sign). If you had already added other files to the repository and made changes across multiple different files (this is pretty typical as your project grows and the changes that you introduce during your work become more complex), different chunks that indicate changes in a different file, would be delineated by headers that look like the fifth line in this output, starting with the "@@" symbols. It is not necessary to run `git diff` if you know exactly what changes you are staging and committing, but it provides a beneficial look into these changes in cases where the changes are complex. In particular, examining the output of `git diff` can give you a hint about what you might write in your commit message. This would be a good point to make a slight aside about these messages and to try to impress upon you that it is a great idea to write commit messages that are clear and informative. This is because as your project changes and the log of your repository grows longer, these messages will serve you (and your collaborators, if more than one person is working with you on the same repository, we will get to that in a little bit) as guide posts to find particular sets of changes that you made along the way. Think of these as messages that you are sending right now, with the memory of the changes you made to the project fresh in your mind, to yourself six months from now when you will no longer remember exactly what you did or why.

At any rate, committing these changes can now look as simple as

```
$ git add my_file.txt
$ git commit -m"Adds a second line of text"
```

which would now increase your log to three different entries:

```
commit f1befd701b2fc09a52005156b333eac79a826a07 (HEAD -> main)
Author: Ariel Rokem <arokem@gmail.com>
Date:    Fri Dec 31 20:26:24 2021 -0800

    Adds a second line of text

commit 42bab7959c3d3c0bce9f753abf76e097bab0d4a8
Author: Ariel Rokem <arokem@gmail.com>
Date:    Fri Dec 31 20:00:05 2021 -0800

    Adds text to the file

commit 1808a8053803150c8a022b844a3257ae192f413c
Author: Ariel Rokem <arokem@gmail.com>
Date:    Fri Dec 31 11:29:31 2021 -0800

    Adds a new file to the repository
```

As you have already seen, this log serves as a record of the changes you made to files in your project. In addition, it can also serve as an entry point to undoing these changes or to see the state of the files as they were at a particular point in time. For example, let's say that you would like to go back to the state that the file had before the addition of the second line of text to that file. To do so, you would use the SHA identifier of the commit that precedes that change and the git checkout command.

```
$ git checkout 42bab my_file.txt
Updated 1 path from f9c56a9
```

Now, you can open the my_file.txt file and see that the contents of this file have been changed to remove the second line of text. Notice also that we did not have to type the entire forty-character SHA identifier. That is because the five first characters uniquely identify this commit within this repository and Git can find that commit with just that information. To reapply the change, you can use the same checkout command, but use the SHA of the later commit instead.

```
$ git checkout f1bef my_file.txt
Updated 1 path from e798023
```

One way to think of the checkout command is that as you are working on your project Git is creating a big library that contains all of the different possible states of your file system. When you want to change your file system to match one of these other states, you can go to this library and check it out from there. Another way to think about this is that we are pointing HEAD to different commits and thereby changing the state of the file system. This capability becomes increasingly useful when we combine it with the use of branches, which we will see next.

3.3 Working with Git at the Second Level: Branching and Merging

As we mentioned before, branches are different states of your project that can exist side by side. One of the dilemmas that we face when we are working on data analysis projects for a duration of time is that we would like to be able to rapidly make experiments and changes to the code, but we often also want to be able to switch over to a state of the code that "just works." This is exactly the way that branches work: using branches allows you to make rapid changes to your project without having to worry that you will not be able to easily recover a more stable state of the project. Let's see this in action in the minimal project that we started working with. To create a new branch, we use the git branch command. To start working on this branch we check it out from the repository (the library) using the git checkout command.

```
$ git branch feature_x
$ git checkout feature_x
```

```
$ echo "this is a line with feature x" >> my_file.txt
$ git add my_file.txt
$ git commit -m"Adds feature x to the file"
```

Examining the git log you will see that there is an additional entry in the history. For brevity, we show here just the two top (most recent) entries, but the other entries will also be in the log (remember that you can leave the log buffer by pressing the q key).

```
commit ab2c28e5c08ca80c9d9fa2abab5d7501147851e1 (HEAD -> feature_x)
Author: Ariel Rokem <arokem@gmail.com>
Date:   Sat Jan 1 13:27:48 2022 -0800

    Adds feature x to the file

commit f1befd701b2fc09a52005156b333eac79a826a07 (main)
Author: Ariel Rokem <arokem@gmail.com>
Date:   Fri Dec 31 20:26:24 2021 -0800

    Adds a second line of text
```

You can also see that HEAD (the current state of the repository) is pointing to the
feature_x branch. We also include the second entry in the log, so that you can see
that the main branch is still in the state that it was before. We can further verify that by
checking out the main branch.

```
$ git checkout main
```

If you open the file now, you will see that the feature x line is nowhere to be seen
and if you look at the log, you will see that this entry is not in the log for this branch and
that HEAD is now pointing to main. But—and this is part of what makes branches so
useful—it is very easy to switch back and forth between any two branches. Another git
checkout feature_x would bring HEAD back to the feature_x branch and
adds back that additional line to the file. In each of these branches, you can continue to
make changes and commit these changes, without affecting the other branch. The history
of the two branches has diverged in two different directions. Eventually, if the changes that
you made on the feature_x branch are useful, you will want to combine the history
of these two branches. This is done using the git merge command, which asks Git to
bring all of the commits from one branch into the history of another branch. The syntax
of this command assumes that HEAD is pointing to the branch into which you would like
to merge the changes, so if we are merging feature_x into main we would issue one
more git checkout main before issuing the merge command.

```
$ git checkout main
$ git merge feature_x
Updating f1befd7..ab2c28e
Fast-forward
my_file.txt | 1 +
1 file changed, 1 insertion(+)
```

The message from Git indicates to us that the main branch has been updated, or fast-
forwarded. It also tells us that this update pulled in changes in one file, with the addition of
one line of insertion. A call to git log will show us that the most recent commit (the
one introducing the feature x line) is now in the history of the main branch and
HEAD is pointing to both main and feature_x.

```
commit ab2c28e5c08ca80c9d9fa2abab5d7501147851e1 (HEAD -> main, feature_x)
Author: Ariel Rokem <arokem@gmail.com>
Date:   Sat Jan 1 13:27:48 2022 -0800

    Adds feature x to the file
```

This is possible because both of these branches now have the same history. If we would
like to continue working, we can remove the feature_x branch and keep going.

```
$ git branch -d feature_x
```

Using branches in this way allows you to rapidly make changes to your projects, while always being sure that you can switch the state of the repository back to a known working state. We recommend that you keep your `main` branch in this known working state and only merge changes from other branches when these changes have matured sufficiently. One way that is commonly used to determine whether a branch is ready to be merged into `main` is by asking a collaborator to look over these changes and review them in detail. This brings us to the most elaborate level of using Git, which is in collaboration with others.

3.4   Working with Git at the Third Level: Collaborating with Others

So far, we have seen how Git can be used to track your work on a project. This is certainly useful when you do it on your own, but Git really starts to shine when you begin to track the work of more than one person on the same project. For you to be able to collaborate with others on a Git-tracked project, you will need to put the Git repository in a place where different computers can access it. This is called a *remote*, because it is a copy of the repository that is on another computer, usually located somewhere remote from your computer. Git supports many different modes of setting up remotes, but we will only show one here, using the GitHub website[5] as the remote.

To get started working with GitHub, you will need to set up a free account. If you are a student or instructor at an educational institution, we recommend that you look into the various educational benefits that are available to you through GitHub Education.[6]

Once you have set up your user account on GitHub, you should be able to create a new repository by clicking on the "+" sign on the top right of the web page and selecting "New repository" (see Fig. 3.1).

On the following page (Fig. 3.2), you will be asked to name your repository. GitHub is a publicly available website, but you can choose whether you would like to make your code publicly viewable (by anyone!) or only viewable by yourself and collaborators that you will designate. You will also be given a few other options, such as an option to add a README file and to add a license. We will come back to these options and their implications when we talk about sharing data analysis code in chapter 7.

The newly created web page will be the landing page for your repository on the internet. It will have a URL that will look something like: `https://github.com/<user name>/<project name>`. For example, since we created this repository under Ariel's GitHub account, it now has the URL `https://github.com/arokem /my_project` (Feel free to visit that URL. It has been designated as publicly viewable). But there is nothing there until we tell Git to transfer over the files from our local

5. https://github.com
6. https://education.github.com/

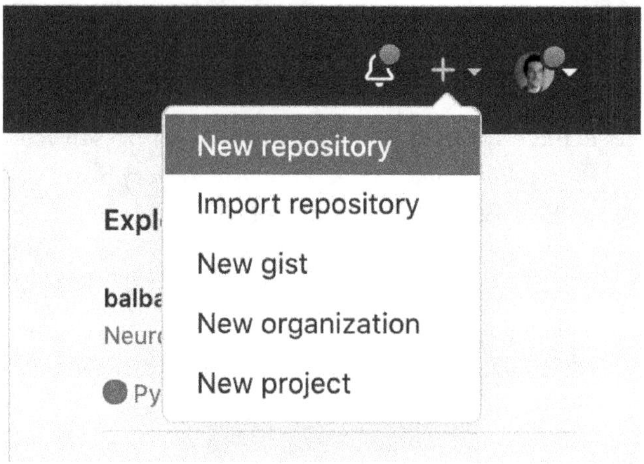

FIGURE 3.1. Creating a new repository through the GitHub web interface.

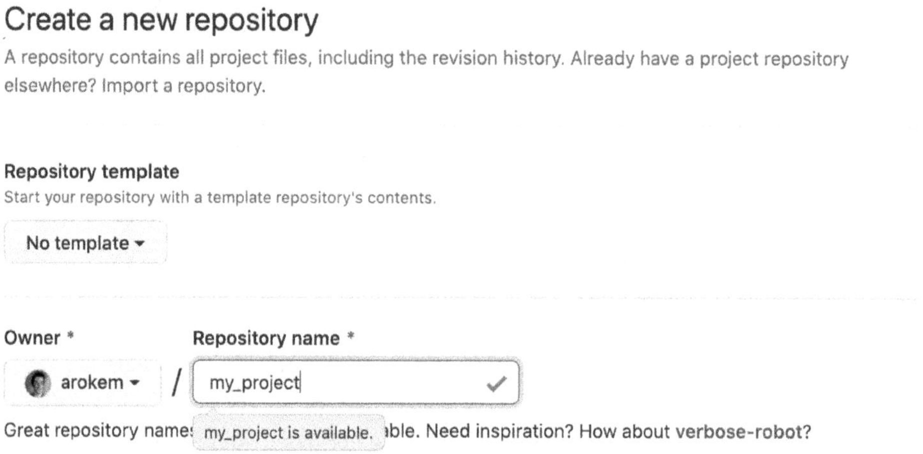

Create a new repository

A repository contains all project files, including the revision history. Already have a project repository elsewhere? Import a repository.

Repository template
Start your repository with a template repository's contents.

No template ▾

Owner *   Repository name *

🔵 arokem ▾   /   my_project   ✓

Great repository name: my_project is available. ble. Need inspiration? How about verbose-robot?

FIGURE 3.2. Initializing a GitHub repository by giving it a name.

copy of the repository on our machine into a copy on the remote, i.e., the GitHub webpage. When a GitHub repository is empty, the front page of the repository contains instructions for adding code to it, either by creating a new repository on the command line or from an existing repository. We will do the latter here because we already have a repository that we have started working on. The instructions for this case are entered in the command line with the shell set to have its working directory in the directory that stores our repository.

```
$ git remote add origin https://github.com/arokem/my_project.git
$ git branch -M main
$ git push -u origin main
```

The first line of these uses the `git remote` subcommand, which manages remotes, and the sub-subcommand `git remote add` to add a new remote called `origin`. We could name it anything we want (e.g., `git remote add github` or `git remote add arokem`), but it is a convention to use `origin` (or sometimes also `upstream`) as the name of the remote that serves as the central node for collaboration. It also tells Git that this remote exists at that URL (notice the addition of `.git` to the URL). Admittedly, it is a long command with many parts.

The next line is something that we already did when we first initialized this repository, namely changing the name of the default branch to `main`. So, we can skip that now.

The third line uses the `git push` subcommand to copy the repository—the files stored in it and its entire history—to the remote. We are telling Git that we would like to copy the `main` branch from our computer to the remote that we have named `origin`. The `-u` flag tells Git that this is going to be our default for `git push` from now on, so whenever we would like to push again from `main` to `origin`, we will only have to type `git push` from now on, and not the entire long version of this incantation (`git push origin main`).

Importantly, though you should have created a password when you created your GitHub account, GitHub does not support authentication with a password from the command line. This makes GitHub repositories very secure, but it also means that when you are asked to authenticate to push your repository to GitHub, you will need to create a personal access token (PAT). The GitHub documentation has a webpage that describes this process.[7] Instead of typing the GitHub password that you created when you created your account, copy this token (typically, a long string of letters, numbers, and other characters) into the shell when prompted for a password. This should look as follows:

```
$ git push -u origin main
Username for 'https://github.com': arokem
Password for 'https://arokem@github.com':<insert your token here>
Enumerating objects: 12, done.
Counting objects: 100% (12/12), done.
Delta compression using up to 8 threads
Compressing objects: 100% (5/5), done.
Writing objects: 100% (12/12), 995 bytes | 497.00 KiB/s, done.
Total 12 (delta 0), reused 0 (delta 0), pack-reused 0
To https://github.com/arokem/my_project.git
* [new branch]      main -> main
Branch 'main' set up to track remote branch 'main' from 'origin'.
```

Once we have done that, we can visit the webpage of the repository again and see that the files in the project and the history of the project have now been copied into that website (Fig. 3.3). Even if you are working on a project on your own this serves as a remote backup, and that is already quite useful to have, but we are now also ready to start collaborating.

---

7. https://docs.github.com/en/authentication/keeping-your-account-and-data-secure/creating-a-personal-access-token

This is done by clicking on the "Settings" tab at the top right part of the repository page, selecting the "Manage access" menu item, and then clicking on the "Add people" button (Fig. 3.4). This will let you select other GitHub users, or enter the email address for your collaborator (Fig. 3.5). As another security measure, after they are added as a collaborator, this other person will have to receive an email from GitHub and approve that they would like to be designated as a collaborator on this repository, by clicking on a link in the email that they received.

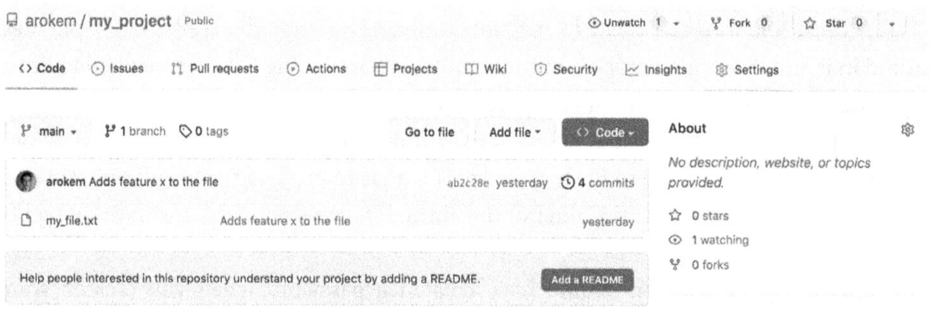

FIGURE 3.3. A copy of the files and history of the project in the GitHub repo.

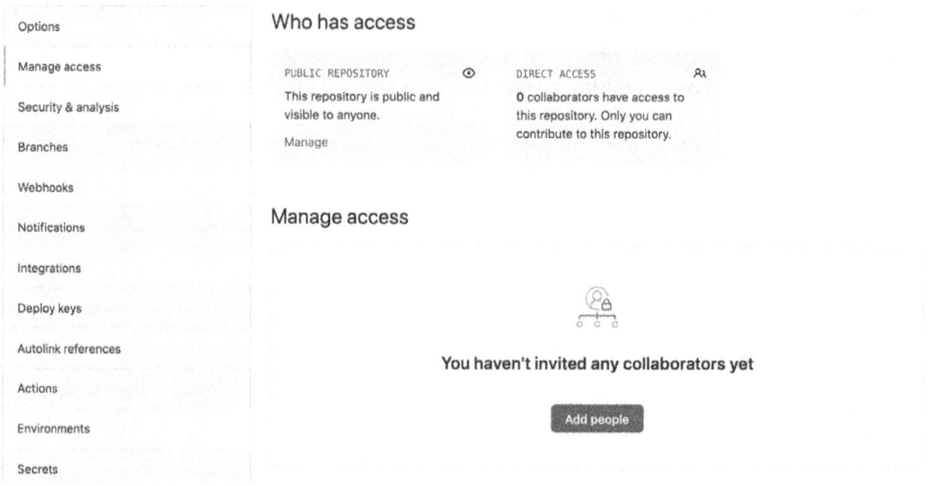

FIGURE 3.4. Click the green "Add people" button to add collaborators.

Once they approve the request to join this GitHub repository as a collaborator, this other person can then both `git push` to your repository, just as we demonstrated previously, as well as `git pull` from this repository, which we will elaborate on subsequently. The first step they will need to take, however, is to `git clone` the repository. This would look something like this:

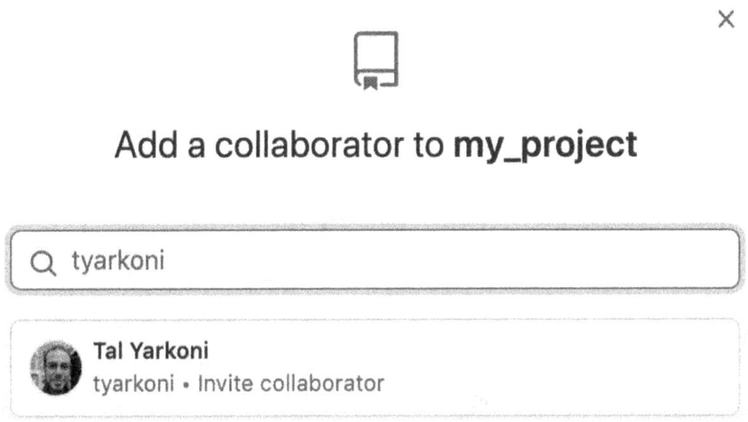

×

Add a collaborator to **my_project**

Q tyarkoni

Tal Yarkoni
tyarkoni • Invite collaborator

FIGURE 3.5. Entering the GitHub username of a collaborator will let you select them from a drop-down.

```
$ git clone https://github.com/arokem/my_project.git
```

When this is executed inside of a particular working directory (e.g., ~/projects/), this will create a subdirectory called my_project (~/projects/my_project; note that you cannot do that in the same directory that already has a my_project folder in it, so if you are following along, issue this git clone command in another part of your file system) that contains all of the files that were pushed to GitHub. In addition, this directory is a full copy of the Git repository, meaning that it also contains the entire history of the project. All of the operations that we did with the local copy on our own repository (e.g., checking out a file using a particular commit's SHA identifier) are now possible on this new copy.

The simplest mode of collaboration is where a single person makes all of the changes to the files. This person is the only person who issues git push commands, which update the copy of the repository on GitHub and then all the other collaborators issue git pull origin main within their local copies to update the files that they have on their machines. The git pull command is the opposite of the git push command that you saw before, syncing the local copy of the repository to the copy that is stored in a particular remote (in this case, the GitHub copy is automatically designated as origin when git clone is issued).

A more typical mode of collaboration is one where different collaborators are all both pulling changes that other collaborators made, as well as pushing their changes. This could work, but if you are working on the same repository at the same time, you might run into some issues. The first arises when someone else pushed to the repository during the time that you were working and the history of the repository on GitHub is out of sync with the history of your local copy of the repository. In that case, you might see a message such as this one:

```
$ git push origin main
To https://github.com/arokem/my_project.git
! [rejected]        main -> main (fetch first)
error: failed to push some refs to 'https://github.com/arokem/my_project.git'
hint: Updates were rejected because the remote contains work that you do
hint: not have locally. This is usually caused by another repository pushing
hint: to the same ref. You may want to first integrate the remote changes
hint: (e.g., 'git pull ...') before pushing again.
hint: See the 'Note about fast-forwards' in 'git push --help' for details.```
```

Often, it would be enough to issue a `git pull origin main`, followed by a `git push origin main` (as suggested in the hint) and this issue would be resolved: your collaborator's changes would be integrated into the history of the repository together with your changes and you would happily continue working. But in some cases, if you and your collaborator introduced changes to the same lines in the same files, Git would not know how to resolve this conflict and you would then see a message that looks like this:

```
$ git pull origin main
remote: Enumerating objects: 5, done.
remote: Counting objects: 100% (5/5), done.
remote: Compressing objects: 100% (2/2), done.
remote: Total 3 (delta 0), reused 3 (delta 0), pack-reused 0
Unpacking objects: 100% (3/3), 312 bytes | 78.00 KiB/s, done.
From https://github.com/arokem/my_project
* branch            main         -> FETCH_HEAD
  ab2c28e..9670d72  main          -> origin/main
Auto-merging my_file.txt
CONFLICT (content): Merge conflict in my_file.txt
Automatic merge failed; fix conflicts and then commit the result.
```

This means that there is a conflict between the changes that were introduced by your collaborator and the changes that you introduced. There is no way for Git to automatically resolve this kind of situation because it cannot tell which of the two conflicting versions should be selected. Instead, Git edits the file to add conflict markers. That is, if you open the text file in a text editor, it would now look something like this:

```
a first line of text
another line of text
this is a line with feature x
<<<<<<< HEAD
an addition you made
=======
addition made by a collaborator
>>>>>>> 9670d72897a5e15defb257010f928bd22f54c929
```

Everything up to the third line of the file is changes that both of you had in your history. After that, your changes to the file diverge. At the same time that you added the line of text an addition you made, the collaborator added the line addition

made by a collaborator. To highlight the source of the conflict, Git introduced the markers you see in the text snippet here. The <<<<<<< HEAD marker
starts the part of the file that was in HEAD on your local copy. Everything below the
======= marker, until the >>>>>>> 9670d72... comes from the collaborator.
The SHA identifier at the end of this marker is also the identifier of the commit that the
collaborator made to the repository while you were making your changes.

In addition, if you issue a git status command at this point, you will see the
following:

```
$ git status
On branch main
Your branch and 'origin/main' have diverged,
and have 1 and 1 different commits each, respectively.
(use "git pull" to merge the remote branch into yours)

You have unmerged paths.
(fix conflicts and run "git commit")
(use "git merge --abort" to abort the merge)

Unmerged paths:
(use "git add <file>..." to mark resolution)
    both modified:   my_file.txt

no changes added to commit (use "git add" and/or "git commit -a")
```

Regardless of whether you would like to include your changes, as well as the changes
from your collaborator, or just one of these change sets, your next step is to edit the files
where conflicts exist to end up with the content that you would want them to have going
forward. This would include removing the conflict markers that Git introduced along the
way. To resolve the conflict, you might also need to talk to your collaborator to understand
why you both made conflicting changes to the same lines. At the end of this process, you
would make one more git commit, followed by a git push, and keep working
from this new state. Your collaborators should be able to issue a git pull in their copy
of the repository to get the conflict resolved on their end as well.

Another slightly more elaborate pattern of collaboration would involve the use of
branches. Remember that we mentioned that maybe it would be a good idea to work only
on branches other than main? One reason for that is that this allows you to go through a
process of review of new changes before they get integrated into the main branch. GitHub
facilitates this process through an interface called a *pull request*. The process works as follows: whenever you would like to make a change to the project, you create a new branch
from main and implement the change on this branch. For example,

```
$ git branch feature_y
$ git checkout feature_y
$ echo "this is a line with feature y" >> my_file.txt
```

Then, we add the file to the staging area, commit this change in this branch, and push the branch to GitHub:

```
$ git add my_file.txt
$ git commit -m"Implements feature y"
[feature_y 948ea1d] Implements feature y
1 file changed, 1 insertion(+)
$ git push origin feature_y
Enumerating objects: 5, done.
Counting objects: 100% (5/5), done.
Delta compression using up to 8 threads
Compressing objects: 100% (2/2), done.
Writing objects: 100% (3/3), 283 bytes | 283.00 KiB/s, done.
Total 3 (delta 1), reused 0 (delta 0), pack-reused 0
remote: Resolving deltas: 100% (1/1), completed with 1 local object.
remote:
remote: Create a pull request for 'feature_y' on GitHub by visiting:
remote:      https://github.com/arokem/my_project/pull/new/feature_y
remote:
To https://github.com/arokem/my_project.git
* [new branch]      feature_y -> feature_y
```

As you can see in the message emitted by Git, a new branch has been created and we can visit the GitHub repository at the URL provided to create a pull request. The page that opens when you visit this URL already includes the commit message that we used to describe these changes. It also has a text box into which we can type more details about the change (Fig. 3.6). Once we hit the "Create pull request" button, we are taken to a separate page that serves as a forum for discussion of these changes (Fig. 3.7). At the same time, all of the collaborators on the repository receive an email that alerts them that a pull request has been created. By clicking on the "Files changed" tab on this page, collaborators can see a detailed view of all of the changes introduced to the code (Fig. 3.8). They can comment on the changes through the web interface and discuss any further changes that

FIGURE 3.6. When opening a pull request, you can add some information about the proposed changes.

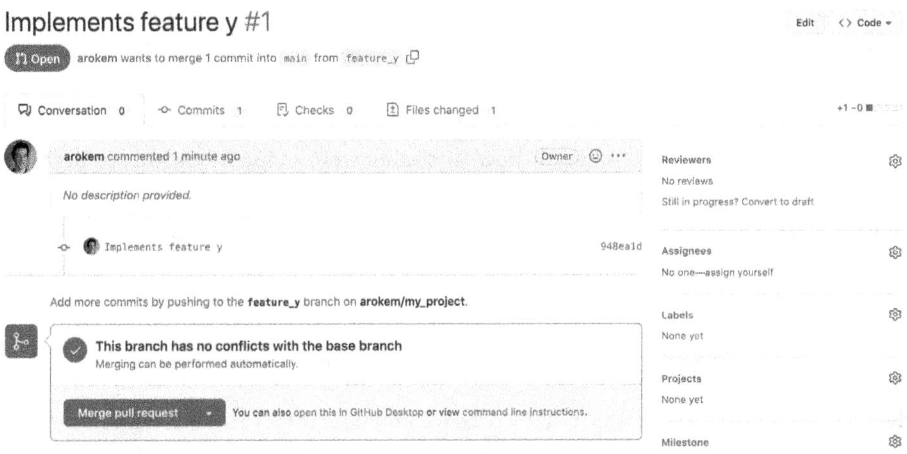

FIGURE 3.7. The pull request interface is a place to discuss proposed changes and eventually incorporate them into the project, by clicking the green "Merge pull request" button.

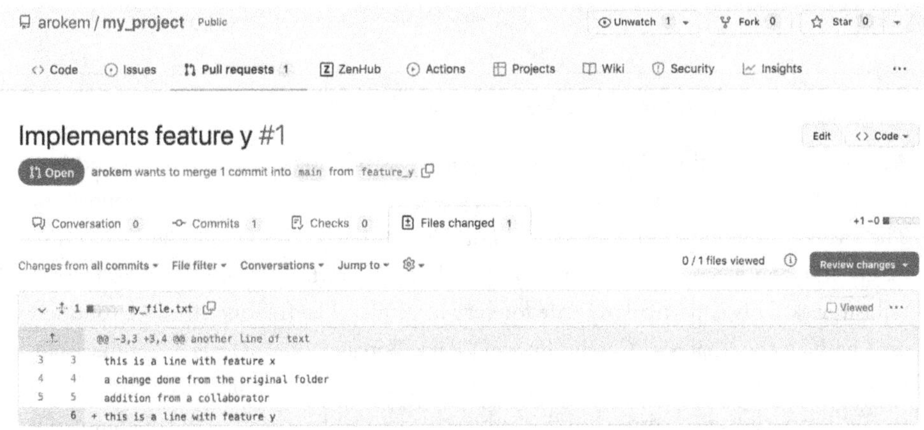

FIGURE 3.8. You can view a detailed line-by-line "diff" of the changes in the "Files changed" tab of a pull request.

need to be made for this contribution to be merged into `main`. The original author of the branch that led to this pull request can continue working on the branch, based on these comments, making further commits. When consensus is reached, one of the collaborators on the repository can click the "Merge pull request" button (Fig. 3.7). This triggers the sending of another email to all of the collaborators on this repository, alerting them that the `main` branch has changed and that they should issue a `git pull origin main` on their local copies of the repository to sync their `main` branch with the GitHub repository.

While this model of collaboration might seem a bit odd at first, it has a lot of advantages: individual contributions to the project are divided into branches and pull requests. The pull request interface allows for a detailed line-by-line review of changes to the project and discussion. This is particularly important for projects where changes to complex code are required. In tandem with automated code testing (which we will discuss in chapter 6), thorough code review mitigates merging changes to the code that introduce new bugs to the code base. In addition to these advantages, the pull request is retained after the change is merged, as a permanent record of these discussions. This allows all of the collaborators on a project to revisit how decisions were made to introduce particular changes. This is a great way to retain collective memory about the way that a project evolved, even after collaborators on the project graduate and/or move on to do other things.

3.4.1   More Complex Collaboration Patterns

Here, we demonstrated a pattern that works well for a relatively small collaboration, with a few collaborators who all know each other. But, Git and GitHub also enable much more complex patterns, that support large and distributed collaborations between people who do not even know each other but would like to contribute to the same project. These patterns enable the large open-source software projects that we will start discussing in chapter 8.

Datalad is Version Control for Data

As you read the section about Git you might be wondering whether it is a good idea to also store the data that you are working with within your project Git repository. The answer is that you probably should not. Git was designed to track changes to text-based files, and it would not operate very well if you started adding data files to your repositories. This is particularly true for very large files, like the ones that we routinely deal with in the analysis of neuroimaging data. Fortunately, there is a Git-like solution for tracking data. This is an open-source software project called Datalad.[8] Similar to Git, it operates as a command-line program that can be called with subcommands to add data to a repository and to track changes to the data. One of the most compelling use-cases for Datalad is in tandem with Git to track changes to code and the changes to derivatives of the code at the same time. We will not go into depth on how to use datalad here, but to learn more about Datalad and to get started using it, you can read an openly available book[9] that introduces Datalad and walks you through its usage.

8. https://www.datalad.org/
9. http://handbook.datalad.org/en/latest/

3.5   Additional Resources

To learn more about the history and more recent objections to the "master/slave" terminology in computer science and engineering, you can refer to [Eglash 2007].

To explain the use of Git and GitHub, we went through a whole sequence of operations using the command line. However, there are several graphical user interfaces (GUIs) for Git and GitHub that are worth checking out. First of all, GitHub has a GUI that they developed.[10] In addition, there is an entire page of options to choose from in the Git documentation.[11]

To learn about more elaborate collaboration patterns for version control and collaboration, we would recommend studying the Git workflow[12] used by the nibabel[13] project. You will learn more about what nibabel does in chapter 12, but for now, you can use their documentation to learn more about how you might use Git.

10. https://desktop.github.com/
11. https://git-scm.com/downloads/guis
12. https://nipy.org/nibabel/gitwash/git_development.html
13. https://nipy.org/nibabel/

4

Computational Environments and Computational Containers

One of the challenges that you will face as you start exploring data science is that doing data science often entails combining many different pieces of software that do different things. It starts with different Python libraries. One of the reasons that people use Python for data science and scientific research, is the many different libraries that have been developed for data science in Python. This is great, but it also means that you will need to install these libraries onto your computer. Fortunately, most Python software is easy to install, but things can get a bit complicated if different projects that you use require different sets of *dependencies*. For example, one project might require one version of a software library, while another project might require another version of the same library. In addition, when working on a project, you often want to move all of these dependencies to a different computer, e.g., to share with a collaborator or a colleague, or so that you can deploy your analysis code to a cluster or a cloud computing system. Here, we will show you two different ways to manage these kinds of situations: *virtual environments* and *containerization*.

4.1   Creating Virtual Environments with Conda

A virtual environment is a directory on your computer that contains all of the software dependencies that are used in one project.

There are a few different ways to set up a working Python installation that will support running scientific software. We recommend the application that we tend to use to install scientific Python libraries and to manage virtual environments, the Conda package manager.[1] To start using that on your system, head over to the Conda installation webpage[2] and follow the instructions for your platform.

If you are working on a Linux or Mac operating system, Conda should become available to you through the shell that you used in the Unix section. If you are working in the

---

1. https://docs.conda.io/en/latest/
2. https://docs.conda.io/projects/conda/en/latest/user-guide/install/index.html

Windows operating system, this will also install another shell into your system (the Conda shell), and you can use that shell for Conda-related operations. We refer you to an online guide from Codeacademy[3] for bringing Conda into the gitbash shell, if you would like to have one unified environment to work in.

After you are done installing Conda, once you start a new terminal shell, your prompt will tell you that you are in Conda's base environment:

```
(base) $
```

Similar to our recommendation never to work in Git's main branch, we also recommend never working in Conda's base environment. Instead, we recommend creating new environments (one for each one of your projects, for example) and installing the software dependencies for each project into the project environment. Similar to Git, Conda also has subcommands that do various things. For example, the conda create command is used to create new environments.

```
$ conda create -n my_env python=3.8
```

This command creates a new environment called my_env (the -n flag signifies that this is the environment's *name*). In this case, we have also explicitly asked that this environment be created with version 3.8 of Python. You can explicitly specify other software dependencies as well. For example,

```
$ conda create -n my_env python=3.8 jupyter
```

would create a new virtual environment that has Python 3.8 and also has the Jupyter notebook software, which we previously mentioned in chapter 1. In the absence of an explicit version number for Jupyter, the most recent version would be installed into this environment. After issuing this command, Conda should ask you to approve a plan for installing the software and its dependencies. If you approve it, and once that is done, the environment is ready to use, but you will also need to activate it to step into this environment.

```
$ conda activate my_env
```

This should change your prompt to indicate that you are now working in this environment:

```
(my_env) $
```

---

3. https://discuss.codecademy.com/t/setting-up-conda-in-git-bash/534473

Once the environment is activated, you can install more software into it, using `conda install`. For example, to install the NumPy software library, which you will learn about in chapter 8, you would issue:

```
(my_env) $ conda install numpy
```

You can run `conda deactivate` to step out of this environment or `conda activate` with the name of another environment to step between environments. To share the details of an environment, or to transfer the environment from one computer to another, you can ask Conda to export a list of all of the software libraries that are installed into the environment, specifying their precise version information. This is done using the `conda env export` command. For example, the following

```
(my_env) $ conda env export > environment.yml
```

exports the details of this environment into a file called `environment.yml`. This uses the YAML[4] markup language—a text-based format—to describe all of the software dependencies that were installed into this environment. You can also use this file to install the same dependencies on another computer on which Conda has been installed by issuing:

```
(base) $ conda create -f environment.yml
```

Because the `environment.yml` file already contains the name of the environment, you do not have to specify the name by passing the `-n` flag. This means that you can replicate the environment on your machine, in terms of Python software dependencies, on another machine. The `environment.yml` file can be sent to collaborators, or you can share your `environment.yml` in the GitHub repo that you use for your project. This is already quite useful, but wouldn't it be nice if your collaborators could get the contents of your computer—Python software dependencies, and also operating system libraries and settings, with the code and also the data—all on their computer with just one command? With containerization, you can get pretty close to that, which is why we will talk about it next.

4.2   Containerization with Docker

Imagine if you could give your collaborators, or anyone else interested in your work, a single command that would make it possible for them to run the code that you ran, in the same way, with the same data, and with all of the software dependencies installed in the same way. Though it is useful, Conda only gets you part of the way there; you can specify a

---

4. https://yaml.org/

recipe to install particular software dependencies and their versions, and Conda does that for you. To get all the way there, we would also need to isolate the operating system, with all of the software that is installed into it, and even data that is saved into the machine. And we would package it for sharing or rapid deployment across different machines. The technology that enables this is called *containerization,* and one of its most popular implementations is called Docker. Like the other applications that you encountered in this chapter, Docker is a command-line interface that you run in the shell.

4.2.1 Getting Started with Docker

To install Docker, we refer you to the most up-to-date instructions on the Docker website.[5] Once installed, you can run it on the command line.[6] Like Git and Conda, Docker also operates through commands and subcommands. For example, the `docker container` command deals with *containers*; i.e., the specific containerized machines that are currently running on your machine (we also refer to your machine in this scenario as the *host* on which the containers are running).

To run a container, you will first have to obtain a Docker *image.* Images are the specification that defines the operating system, the software dependencies, programs, and even data that are installed into a container that you run. So, you can think of the image as a computational blueprint for producing multiple containers that are all identical to each other, or the original footage of a movie, which can be copied, edited, and played on multiple different devices.

To get an image, you can issue the `docker pull` command:

```
$ docker pull hello-world
Using default tag: latest
latest: Pulling from library/hello-world
2db29710123e: Pull complete
Digest: sha256:09ca85924b43f7a86a14e1a7fb12aadd75355dea7cdc64c9c80fb2961cc53fe7
Status: Downloaded newer image for hello-world:latest
docker.io/library/hello-world:latest
```

This is very similar to the `git pull` command. Per default, Docker looks for the image in the dockerhub registry,[7] but you can also ask it to pull from other registries (which is the name for a collection of images). Docker tells you that it is pulling the image, and which version of the image was pulled. Much like Git commits in a repo, Docker images are identified through a secure hash algorithm (SHA) identifier. In addition, because it is very important to make sure that you know exactly which version of the image you are using, images can be labeled with a *tag.* The most recent version of the image

5. https://docs.docker.com/get-docker/

6. In some cases, you might also need to create an account on DockerHub at https://hub.docker.com and log into this account through the application that was installed on your computer when you installed the Docker command-line interface.

7. https://hub.docker.com

that was pushed into the registry from which you are pulling always has the tag `latest` (which means that this tag points to different versions of the image at different times, so you have to be careful interpreting it!). Once you have pulled it, you can run this image as a container on your machine, and you should see the following text, which tells you that this image was created mostly so that you can verify that running images as containers on your machine works as expected.

```
$ docker run hello-world

Hello from Docker!
This message shows that your installation appears to be working correctly.

To generate this message, Docker took the following steps:
1. The Docker client contacted the Docker daemon.
2. The Docker daemon pulled the "hello-world" image from the Docker Hub.
   (amd64)
3. The Docker daemon created a new container from that image which runs the
   executable that produces the output you are currently reading.
4. The Docker daemon streamed that output to the Docker client, which sent it
   to your terminal.

To try something more ambitious, you can run an Ubuntu container with:
$ docker run -it ubuntu bash

Share images, automate workflows, and more with a free Docker ID:
https://hub.docker.com/

For more examples and ideas, visit:
https://docs.docker.com/get-started/
```

Once it has printed this message, however, this container will stop running (that is, it is not persistent), and you will be back in the shell prompt in the host machine.

Next, as suggested in the hello world message, we can try to run a container with the image of the Ubuntu operating system, a variety of Linux. The suggested command includes two command line flags. The `i` flag means that the Docker image will be run interactively and the `t` flag means that the image will be run as a terminal application. We also need to tell it exactly which variety of terminal we would like to run. The suggested command ends with a call to the `bash` Unix command, which activates a shell called Bash. This is a popular variety of the Unix shell, which includes the commands that you saw before in chapter 2. This means that when we run this command, the shell on our machine will drop us into a shell that is running inside the container.

```
$ docker run -it ubuntu
Unable to find image 'ubuntu:latest' locally
latest: Pulling from library/ubuntu
ea362f368469: Already exists
Digest: sha256:b5a61709a9a44284d88fb12e5c48db0409cfad5b69d4ff8224077c57302df9cf
Status: Downloaded newer image for ubuntu:latest
root@06d473bd6732:/#
```

The first few lines indicate that Docker identifies that it does not have a copy of the Ubuntu image available locally on the machine, so it automatically issues the `docker pull` command to get it. After it is done pulling, the last line you see in the output is the prompt of the shell inside the running Docker container. You can execute various Unix commands here, including those that you have seen previously. For example, try running `pwd`, `cd`, and `ls` to see where you are and what is in there. Once you are ready to return to your machine, you can execute the `exit` command, which will return you to where you were in your host operating system.

This becomes more useful when you can run some specific application within the container. For example, there is a collection of images that will run the Jupyter notebook in a container. Depending on the specific image that you run, the container will also have other components installed. For example, the `jupyter/scipy-notebook` image includes components of the scientific Python ecosystem (you will learn more about these starting in chapter 8). So executing the following will start the Jupyter notebook application. Because Jupyter runs as a web application of sorts, we need to set up one more thing, which is that it will need to communicate with the browser through a *port*—an identified operating system process that is in charge of this communication. However, because the application is isolated inside the container, if we want to access the port on which the application is running from the host machine, we need to forward the port from the container to the host. This is done using the `-p` flag, as follows,

```
$ docker run -p 8888:8888 jupyter/scipy-notebook
```

where the `8888:8888` input to this flag means that the port numbered 8888 in the container (to the right of the colon) is forwarded to port 8888 in the host (to the left of the colon). When you run this you will first see Docker pulling the `latest` version of this image and then the outputs of Jupyter running in the container. In the end, you will see a message that looks something like this:

```
To access the notebook, open this file in a browser:
    file:///home/jovyan/.local/share/jupyter/runtime/nbserver-7-open.html
Or copy and paste one of these URLs:
    http://dd3dc49a5f0d:8888/?token=a90483637e65d99966f61a2b6e87d447cea6504509cfbefc
 or http://127.0.0.1:8888/?token=a90483637e65d99966f61a2b6e87d447cea6504509cfbefc
```

Copying the last of these URLs (the one that starts with `http://127.0.0.1:8888`) into your browser URL bar should open the Jupyter application in your browser and selecting the "Python 3 (ipykernel)" option would then open a Jupyter notebook in your browser, ready to import all of the components that are installed into the container.

Importantly, if you save anything in this scenario (e.g., new notebook files that you create) it will be saved into the file system of the container. That means that as soon as the container is stopped, these files will be deleted. To avoid this, similar to the mapping of the port in the container to a port on the host, you can also map a location in your file system to the container file system, using the -v flag. For example, the following command would mount Ariel's projects directory (to the left of the colon) to the location /home/jovyan/work inside the container (to the right of the colon).

```
$ docker run -p 8888:8888 -v /Users/arokem/projects/:/home/jovyan/work
↪  jupyter/scipy-notebook
```

More generally, you can use the following command to mount your present working directory into that location:

```
$ docker run -p 8888:8888 -v $(pwd):/home/jovyan/work jupyter/scipy-notebook
```

If you do that, when you direct your browser to the URL provided in the output and click on the directory called work in the notebook interface, you will see the contents of the directory from which you launched Docker, and if you save new notebook files to that location, they will be saved in this location on your host, even after the container is stopped.

4.2.2 Creating a Docker Container

One of the things that makes Docker so powerful is the fact that images can be created using a recipe that you can write down. These files are always called Dockerfile, and they follow a simple syntax, which we will demonstrate next. Here is an example of the entire contents of a Dockerfile:

```
FROM jupyter/scipy-notebook:2022-01-14
RUN conda install -y nibabel
```

The first line in this Dockerfile tells Docker to base this image on the scipy-notebook image. The all-caps word FROM is recognized as a Docker command. In particular, on the 2022-01-14 tag of this image. This means that everything that was installed in that tag of the `scipy-notebook` Docker image will also be available in the image that will be created based on this `Dockerfile`. On the second line, the word RUN is also a Docker command, and anything that comes after it will be run as a Unix command. In this case, the command we issue instructs Conda to install the NiBabel Python library into the image. For now, all you need to know about NiBabel is that it is a neuroimaging library that is not installed into the `scipy-notebook` image (and that you will learn more about it in chapter 12). The `-y` flag is here to indicate that Conda will not need to ask for confirmation of the plan to install the various dependencies, which it does per default.

In the working directory in which the Dockerfile file is saved, we can issue the `docker build` command. For example:

```
$ docker build -t arokem/nibabel-notebook:0.1 .
```

The final argument in this command is a (“.”), which indicates to Docker that it should be looking within the present working directory for a `Dockerfile` that contains the recipe to build the image. In this case the `-t` flag is a flag for naming and tagging the image. It is very common to name images according to the construction `<dockerhub username>/<image name>:<tag>`. In this case, we asked Docker to build an image called `arokem/nibabel-notebook` that is tagged with the tag `0.1`, perhaps indicating that this is the first version of this image. This image can then be run in a container in much the same way that we ran the `scipy-notebook` image in the earlier example, except that we now also indicate the version of this container.

```
$ docker run -p 8888:8888 -v $(pwd):/home/jovyan/work arokem/nibabel-notebook:0.1
```

This produces an output that looks a lot like the one we saw when we ran the `scipy-notebook` image, but importantly, this new container also has NiBabel installed into it, so we have augmented the container with more software dependencies. If you would like to also add data into the image, you can amend the Dockerfile as follows.

```
FROM jupyter/scipy-notebook:2022-01-14
RUN conda install -y nibabel
COPY data.nii.gz data.nii.gz
```

The COPY command does what the name suggests, copying from a path in the host machine (the first argument) to a path inside the image. Let's say that we build this image and tag it as `0.2`:

```
$ docker build -t arokem/nibabel-notebook:0.2 .
```

We can then run this image in a container using the same command that we used previously, changing only the tag:

```
$ docker run -p 8888:8888 -v $(pwd):/home/jovyan/work arokem/nibabel-notebook:0.2
```

Now, when we start the Jupyter application in our web browser, we see that the data file is already placed into the top-level directory in the container:

Importantly, this file is there not because the host file system is being mounted into the machine with the -v flag, but because we copied it into the image. This means that if we shared this image with others, they would also have that same data file in the container that they run with this image. Fortunately, Docker makes it really straightforward to share images, which we will see next.

4.2.3 Sharing Docker Images

Once you have created a Docker image you can share it with others by pushing it to a container registry, using the `docker push` command.

```
$ docker push arokem/nibabel-notebook:0.2
```

If we do not tell Docker otherwise, it will push this to the container registry in Docker-Hub. The same one from which we previously pulled the `hello-world` and also the `scipy-notebook` images. This also means that you can see the Docker image that we created here on the DockerHub website.[8] After you push your image to DockerHub or any other publicly available registry, other people will be able to pull it using the `docker pull` command; however, if you followed the sequence we used here, they will also need to explicitly point both to your DockerHub username and the tag they would like to pull. For example:

```
$ docker pull arokem/nibabel-notebook:0.2
```

---

8. https://hub.docker.com/repository/docker/arokem/nibabel-notebook

> ## NeuroDocker Is Docker for Neuroscience
>
> Outside of the Python software libraries, installing neuroscience software into a Docker image can be quite complicated. Making this easier is the goal of the NeuroDocker project.[9] This project is also a command-line interface that can generate dockerfiles for the creation of Docker containers that have installed in them various neuroimaging software, such as Freesurfer,[10] AFNI,[11] ANTS,[12] and FSL.[13] If you use any of these in your analysis pipelines and you would like to replicate your analysis environment across computers, or share it with others, NeuroDocker is the way to go.
>
> 9. https://www.repronim.org/neurodocker/
> 10. https://surfer.nmr.mgh.harvard.edu/
> 11. https://afni.nimh.nih.gov/
> 12. https://picsl.upenn.edu/software/ants/
> 13. https://fsl.fmrib.ox.ac.uk/fsl/fslwiki

4.3   Setting Up

Now, that you have learned a bit about the data science environment, we are ready to set up your computer so that you can follow along and run the code in the chapters that follow. There are several options on how to do this and up-to-date instructions on installing all of the software should always be available on the book website.[14]

Here, we will focus on the simplest method, which uses Docker. Using what you have learned previously, this should be straightforward

```
$ docker pull ghcr.io/neuroimaging-data-science/
↪  neuroimaging-data-science:latest
$ docker run -p 8888:8888 ghcr.io/neuroimaging-data-science/
↪  neuroimaging-data-science:latest
```

After running this, you should see some output in your terminal, ending with a URL that starts with `http://127.0.0.1:8888/?token=`. Copy the entire URL, including the token value which follows, into your browser URL bar. This should bring up the Jupyter notebook interface, with a directory called "contents," which contains the notebooks with all of the content of the book. Note that these notebooks are stored in the markdown format `.md`, but you should still be able to run them as intended.[15] You should be able to run through each of the notebooks (using shift+enter to execute each cell) as you follow along the contents of the chapters. You should also be able to make changes to the contents of the notebooks and rerun them with variations to the code, but be sure to download or copy elsewhere the changes that you make (or to use the `-v` flag we showed

14. http://neuroimaging-data-science.org

15. This works because we rely on a software library called jupytext to translate the markdown text into a format that jupyter knows how to run. See https://jupytext.readthedocs.io/en/latest/

above to mount a host file system location and copying the saved files into this location), because these files will be deleted as soon as the container is stopped.

4.4   Additional Resources

Other virtual environment systems for Python include `venv`.[16] This is less focused on scientific computing, but is also quite popular.

If you are looking for options for distributing containers outside of DockerHub, one option is the GitHub container registry,[17] which integrates well with software projects that are tracked on GitHub.

---

16. https://docs.python.org/3/library/venv.html

17. https://docs.github.com/en/packages/working-with-a-github-packages-registry/working-with-the-container-registry

PART II

Programming

5

A Brief Introduction to Python

Researchers in neuroimaging use different programming languages to perform data analysis. In this book, we chose to focus on the Python programming language. However, as we mentioned in chapter 1, this book is not meant to be a general introduction to programming; out of necessity, we are going to assume that you have had some prior experience writing code in one or more other programming languages (or possibly even Python itself). That said, we will be *briefly* reviewing key programming concepts as we go, so you do not need to have had *much* prior experience. As long as you have encountered variables, functions, and for-loops before, you should be just fine. And if you have not yet, we recommend some resources at the end of this section to get you up to speed. Conversely, if you have been programming in other languages for years, you should be able to breeze through this section very quickly, though we might still recommend paying attention to the last few subsections, which talk about some deeper ideas underlying the Python language.

5.1   What Is Python?

Let's start by talking a bit about what Python is and why we chose it for this book. First of all, it is a programming language, which means it is a set of rules for writing instructions that a computer can understand and follow. Of course, that alone does not make Python special; there are hundreds of other programming languages out there! But Python is not just *any* programming language; by many rankings, it is currently (as we write this in 2023) the world's single most popular language. And it is particularly dominant in one of this book's two core areas of focus: data science. We also happen to think it is by far the best choice for doing serious work in the book's other core area of focus, neuroimaging, and it is the language that we use most often for our neuroimaging work. But you do not have to take our word for that right now; hopefully, you will be convinced of it as we work through this book.

Why do so many people like Python? There are many answers to that, but here are a few important ones.

First, Python is a *high-level, interpreted* programming language. This means that, in contrast to low-level, compiled languages like C, C++, Java, etc., Python features a high level

55

of abstraction. A lot of the things you have to worry about in many low-level languages (e.g., memory allocation, garbage collection) are done for you automatically in Python.

Second, Python's syntax is readable and (relatively speaking) easy to learn. As you will see shortly, many Python operators are ordinary English words. Python imposes certain rules on code structure that most languages do not. This may be a bit annoying at first, but it also makes it easier to read other people's code once you acclimate. One of the consequences of these design considerations is that mathematical ideas translate into code in a way that does not obscure the math. That means that mathematical equations implemented in Python look a lot like the same equations written on paper. This is useful in data science applications, where ideas are expressed in mathematical terms.

Third, Python is a *general-purpose* language. In contrast to many other dynamic programming languages designed to serve specific niche uses, Python is well-suited for a wide range of applications. It features a comprehensive standard library (i.e., the functionality available out-of-the-box when you install Python) and an enormous ecosystem of third-party packages. It also supports multiple programming paradigms to varying extents (object oriented, functional, etc.). Consequently, Python is used in many areas of software development. Very few other languages can boast that they have some of the best libraries implemented in *any* language for tasks as diverse as, say, scientific computing and back-end web development.

Lastly, there is the sheer size of the Python community. While Python undeniably has many attractive features, we would not want to argue that it is a *better* overall programming language than anything else. To some degree, it is probably true that Python's popularity is an accident of history. If we could randomly rerun the last two decades, we might be extolling the virtues of Haskell (or Julia, or Ruby, or...) instead of Python. So we are not saying Python is *intrinsically* the world's greatest language. But there is no denying that there are immense benefits to using a language that so many other people use. For most tasks you might want to undertake if you are doing data science and/or neuroimaging data analysis, finding good libraries, documentation, help, and collaborators is simply going to be much easier if you work in Python than if you work in almost any other language.

Importantly, the set of tools in Python specifically for the analysis of neuroimaging data have rapidly evolved and matured in the last couple of decades, and they have gained substantial popularity through their use in research (we will dive into these tools in part 4 of the book). The neuroimaging in Python ecosystem also has a strong ethos of producing high-quality open-source software, which means that the Python tools that we will describe in this book should be accessible to anyone. More broadly, Python has been adopted across many different research fields and is one of the most commonly used programming languages in the analysis of data across a broad range of domains, including astronomy, geosciences, natural language processing, and so on. For researchers who are thinking of applying their skills in the industry outside of academia, Python is very popular in the industry, with many entry positions in industry data science specifically mentioning experience programming in Python as a desired skill.

With that basic sales pitch out of the way, let's dive right into the Python language. We will spend the rest of this chapter working through core programming concepts and showing you how they are implemented in Python. It should go without saying that this can only hope to be a cursory overview; there just is not time and space to provide a full introduction to the language! But we will introduce many more features and concepts throughout the rest of the book, and we also end each chapter with a curated list of additional resources you can look into if you want to learn more.

5.2 Variables and Basic Types

It is common to introduce programming by first talking about variables, and we will not break with this convention. A variable, as you probably already know, is a store of data that can take on different values (hence the name *variable*), and is typically associated with a fixed name.

5.2.1 Declaring Variables

In Python, we declare a variable by writing its name and then assigning it a value with the equal (=) sign:

```
my_favorite_variable = 3
```

Notice that when we initialize a variable, we do not declare its *type* anywhere. If you are familiar with *statically typed* languages like C++ or Java, you are probably used to having to specify what type of data a variable holds when you create it. For example, you might write `int my_favorite_number = 3` to indicate that the variable is an integer. In Python, we do not need to do this. Python is *dynamically typed*, meaning that the type of each variable will be determined on the fly, once we start executing our program. It also means we can change the type of variable on the fly, without anything bad happening. For example, you can overwrite it with a character string value instead of the integer value that was once stored in this variable:

```
my_favorite_variable = "zzzzzzz"
```

5.2.2 Printing Variables

We can examine the contents of a variable at any time using the built-in `print()` function:

```
print(my_favorite_variable)
```

```
zzzzzzz
```

If we are working in an interactive environment like a Jupyter notebook, we may not even need to call `print()`, as we will automatically get the output of the last line evaluated by the Python interpreter:

```
# This line will not be printed because it is not the
# last line of the notebook cell to be evaluated.
"this line won't be printed"

# but this one will
my_favorite_variable
```

```
'zzzzzzz'
```

As you can see in this example, the hash sign (#) is used for comments. That means that any text that is after a # is ignored by the Python interpreter and not evaluated.

5.2.3 Built-In Types

All general-purpose programming languages provide the programmer with different *types* of variables—things like strings, integers, booleans, and so on. These are the most basic building blocks a program is made up of. Python is no different and provides us with several built-in types.[1] Let's take a quick look at some of these.

INTEGERS

An integer is a numerical data type that can only take on finite whole numbers as its value. For example:

```
number_of_subjects = 20
number_of_timepoints = 1000
number_of_scans = 10
```

Any time we see a number written somewhere in Python code, and it is composed only of digits (no decimals, quotes, etc.), we know we are dealing with an integer.

In Python, integers support all of the standard arithmetic operators you are familiar with: addition, subtraction, multiplication, etc. For example, we can multiply the two variables we just defined

```
number_of_subjects * number_of_timepoints
```

```
20000
```

---

1. https://docs.python.org/3/library/stdtypes.html

or divide one integer by another:

```
number_of_timepoints / number_of_scans
```

```
100.0
```

Notice that the result of the previous division is *not* itself an integer! The decimal point in the result gives away that the result is of a different type: a *float*.

FLOATS

A float (short for *floating point number*) is a numerical data type used to represent real numbers. As we just saw, floats are identified in Python by the presence of a decimal.

```
roughly_pi = 3.14
mean_participant_age = 24.201843727
```

All of the standard arithmetic operators work on floats just like they do on `int`s:

```
print(roughly_pi * 2)
```

```
6.28
```

We can also freely combine `int`s and `float`s in most operations:

```
print(0.001 * 10000 + 1)
```

```
11.0
```

Observe that the output is of type `float`, even though the value is a whole number, and hence could in principle have been stored as an `int` without any loss of information. This is a general rule in Python: arithmetic operations involving a mix of `int` and `float` operands will almost always return a `float`. Some operations will return a `float` even if all operands are `int`s, as we saw previously in the case of division.

Exercise

The Python built-in `type()` function reports to you the type of a variable that is passed to it. Use the `type` function to verify that `number_of_subjects * number_of_timepoints` is a Python integer, while `number_of_timepoints / number_of_scans` is not. Why do you think that Python changes the result of division into a variable of type float?

STRINGS

A string is a sequence of characters. In Python, we define strings by enclosing zero or more characters inside a pair of quotes (either single or double quotes work equally well, so you can use whichever you prefer; just make sure the opening and closing quotes match).

```
country = "Madagascar"
ex_planet = 'Pluto'
```

Python has very rich built-in functionality for working with strings. Let's look at some of the things we can do.

We can calculate the length of a string:

```
len(country)
```

```
10
```

Or we can convert it to uppercase (also try `.lower()` and `.capitalize()`):

```
country.upper()
```

```
'MADAGASCAR'
```

We can count the number of occurrences of a substring (in this case, a single letter `a`):

```
country.count("a")
```

```
4
```

Or we can replace a matching substring with another substring:

```
country.replace("car", "truck")
```

```
'Madagastruck'
```

One thing that you might notice in the previous examples is that they seem to use two different syntaxes. In the first example, it looks like `len()` is a *function* that takes a string as its parameter (or *argument*). By contrast, the last three examples use a different *dot* notation, where the function comes after the string (as in `country.upper()`). If you find this puzzling, do not worry! We will talk about the distinction in much more detail subsequently.

Exercise

Write code to count how many times the combination "li" appears in the string "super-califragilisticexpialidocious." Assign this value into a new variable named `number_of_li` and print its value.

BOOLEANS

Booleans operate pretty much the same in Python as in other languages; the main thing to recognize is that they can only take on the values `True` or `False`. Not `true` or `false`, not `"true"` or `"false."` The only values a boolean can take on in Python are `True` and `False`, written exactly that way. For example:

```
enjoying_book = True
```

One of the ways that Boolean values are typically generated in Python programs is through logical or comparison operations. For example, we can ask whether the length of a given string is greater than a particular integer

```
is_longer_than_2 = len("apple") > 2
print(is_longer_than_2)
```

```
True
```

or whether the product of the first two numbers below equals the third.

```
is_the_product = 719 * 1.0002 == 2000
print(is_the_product)
```

```
False
```

Or we might want to know whether the conjunction of several subexpressions is `True` or `False`:

```
("car" in country) and (len("apple") > 2) and (15 / 2 > 7)
```

```
True
```

As simple as it is, this last example illustrates a nice feature of Python: its syntax is more readable than that of most other programming languages. In it we ask if the substring `"car"` is contained in the string `country` using the English language word in. Similarly, Python's logical conjunction operator is the English word and. This means that we can often quickly figure out—or at least, vaguely intuit—what a piece of Python code does.

Exercise

Some integer values are equivalent to the Python Boolean values. Use the equality (==) operator to find integers that are equivalent to True and that are equivalent to False.

NONE

In addition to these usual suspects, Python also has a type called None. None is special and indicates that no value has been assigned to a variable. It is roughly equivalent to the null value found in many other languages.

```
name = None
```

Note: None and False are *not* the same thing!

```
name == False
```

```
False
```

Also, assigning the value None to a variable is not equivalent to not defining the variable in the first place. Instead, a variable that is set to None is something that we can point to in our program without raising an error but does not carry any particular value. These are subtle but important points, and in later chapters, we will use code where the difference becomes pertinent.

5.3 Collections

Most of the code we will write in Python will require more than just integers, floats, strings, and Booleans. We are going to need more complex data structures, or *collections*, that can hold other objects (like strings, integers, etc.) and enable us to easily manipulate them in various ways. Python provides built-in support for many common data structures, and others can be found in modules that come installed together with the language itself; that is, the so-called *standard library* (e.g., in the collections[2] module).

5.3.1 Lists

Lists are the most common collection we will work with in Python. A list is a *heterogeneous* collection of objects. By heterogeneous, we mean that a list can contain elements of

---

2. https://docs.python.org/3/library/collections.html

different types. It does not have to contain only strings or only integers; it can contain a mix of the two, as well as all kinds of other types.

LIST INITIALIZATION

To create a new list, we enclose one or more values between square brackets ([and]). Elements are separated by commas. Here is how we initialize a list containing four elements of different types (an integer, a float, and two strings).

```
random_stuff = [11, "apple", 7.14, "banana"]
```

LIST INDEXING

Lists are *ordered* collections, by which we mean that a list retains a memory of the position each of its elements was inserted in. The order of elements will not change unless we explicitly change it. This allows us to access individual elements in the list directly, by specifying their position in the collection, or *index*.

To access the *i*th element in a list, we enclose the index *i* in square brackets. Note that Python uses zero-based indexing (i.e., the first element in the sequence has index 0), and not one as in some other data-centric languages (Julia, R, etc.). For example, it means that the following operation returns the second item in the list and not the first.

```
random_stuff[1]
```

```
'apple'
```

Many bitter wars have been fought on the internet over whether zero-based or one-based indexing is better. We are not here to take a philosophical stand on this issue; the fact of the matter is that Python indexing is zero-based, and that is not going to change. So whether you like it or not, you will need to make your peace with the idea that indexing starts from zero while you are reading this book.

Exercise

In addition to indexing from the beginning of the list, we can index from the end of the list using negative numbers (e.g., random_stuff[-1]). Experiment indexing into the list random_stuff with negative numbers. What is the negative number index for the last item in the list? What is the negative number index for the first item in the list? Can you write code that would use a negative number to index the first item in the list, without advance knowledge of its length?

Indexing is nice, but what if we want to pull more than one element at a time out of our list? Can we easily retrieve only part of a list? The answer is yes! We can *slice* a list, and get back another list that contains multiple contiguous elements of the original list, using the colon (:) operator.

```
random_stuff[1:3]
```

```
['apple', 7.14]
```

In the list-slicing syntax, the number before the colon indicates the start position, and the number after the colon indicates the end position. Note that the start is inclusive and the end is exclusive. That is, in the previous example, we get back the second and third elements in the list, but *not* the fourth. If it helps, you can read the 1 : 3 syntax as saying "I want all the elements in the list starting at index 1 and stopping just before index 3."

Lists are *mutable* objects, meaning that they can be modified after they have been created. In particular, we very often want to replace a particular list value with a different value. To overwrite an element at a given index we assign a value to it using the same indexing syntax we saw previously:

```
print("Value of first element before re-assignment:", random_stuff[0])

random_stuff[0] = "eleventy"

print("Value of first element after re-assignment:", random_stuff[0])
```

```
Value of first element before re-assignment: 11
Value of first element after re-assignment: eleventy
```

It is also very common to keep appending variables to an ever-growing list. We can add a single element to a list via the .append() function (notice again that we are calling a function using the dot notation; we promise that we will come back to that later).

```
random_stuff.append(88)
print(random_stuff)
```

```
['eleventy', 'apple', 7.14, 'banana', 88]
```

Exercise

There are several ways to combine lists, including the `append` function you saw in the previous example, as well as the `extend` method. You can also add lists together using the addition (+) operator.

Given the following two lists:

```
list1 = [1, 2, 3]
list2 = [4, 5, 6]
```

How would you create a new list called `list3` that has the items: `[6, 5, 1, 2, 3]`, with as few operations as possible and only using indexing operations and functions associated with the list (Hint: you can look up these functions in the Python online documentation for lists[3])?

---

3. https://docs.python.org/3/tutorial/datastructures.html#more-on-lists

---

5.3.2 Dictionaries

Dictionaries are another extremely common data structure in Python. A dictionary (or `dict`) is a mapping from keys to values; we can think of it as a set of key/value pairs, where the keys have to be unique (but the values do not). Many other languages have structures analogous to Python's dictionaries, though they are usually called something like *associative arrays*, *hash tables*, or *maps*.

DICTIONARY INITIALIZATION

We initialize a dictionary by specifying comma-delimited key/value pairs inside curly braces. Keys and values are separated by a colon. It looks like this:

```
fruit_prices = {
    "apple": 0.65,
    "mango": 1.5,
    "strawberry": "$3/lb",
    "durian": "unavailable",
    5: "just to make a point"
}
```

Notice that both the keys and the values can be heterogeneously typed (observe the last pair, where the key is an integer).

ACCESSING VALUES IN A DICTIONARY

In contrast to lists, you cannot access values stored in a dictionary directly by their serial position. Instead, values in a dictionary are accessed by their key. The syntax is identical to that used for list indexing. We specify the key whose corresponding value we would like to retrieve in between square brackets:

```
fruit_prices['mango']
```

```
1.5
```

And again, the following example would fail, raising a `KeyError` telling us there is no such key in the dictionary:

```
fruit_prices[0]
```

```
--------------------------------------------------------------------
KeyError                                      Traceback (most recent call last)
Input In [30], in <cell line: 1>()
----> 1 fruit_prices[0]

KeyError: 0
```

However, the reason the previous key failed is *not* that integers are invalid keys. To prove that, consider the following:

```
fruit_prices[5]
```

```
'just to make a point'
```

Superficially, it might look like we are requesting the sixth element in the dictionary and getting back a valid value. But that is not what is actually happening here. If it is not clear to you why `fruit_prices[0]` fails while `fruit_prices[5]` succeeds, go back and look at the code we used to create the `fruit_prices` dictionary. Carefully inspect the keys and make sure you understand what is going on.

UPDATING A DICTIONARY

Updating a dictionary uses the same `[]`-based syntax as accessing values, except we now make an explicit assignment. For example, we can add a new entry for the `ananas` key,

```
fruit_prices["ananas"] = 0.5
```

or overwrite the value for the `mango` key

```
fruit_prices["mango"] = 2.25
```

and then look at the dict again.

```
print(fruit_prices)
```

```
{'apple': 0.65, 'mango': 2.25, 'strawberry': '$3/lb', 'durian': 'unavailable', 5:
 'just to make a point', 'ananas': 0.5}
```

Exercise

Add another fruit to the dictionary. This fruit should have several different values associated with it, organized as a list. How can you access the second item in this list in one single call?

5.3.3 Tuples

The last widely used Python collection we will discuss here (though there are many other more esoteric ones) is the *tuple*. Tuples are very similar to lists in Python. The main difference between lists and tuples is that lists are *mutable*, meaning, they can change after initialization. Tuples are *immutable*; once a tuple has been created, it can no longer be modified.

We initialize a tuple in much the same way as a list, except we use parentheses (round brackets) instead of square brackets:

```
my_tuple = ("a", 12, 4.4)
```

Just to drive home the immutability of tuples, let's try replacing a value and see what happens:

```
my_tuple[1] = 999
```

```
---------------------------------------------------------------------
TypeError                                 Traceback (most recent call last)
Input In [36], in <cell line: 1>()
----> 1 my_tuple[1] = 999

TypeError: 'tuple' object does not support item assignment
```

Our attempt to modify the tuple raises an error. Fortunately, we can easily convert any tuple to a list, after which we can modify it to our heart's content.

```
converted_from_tuple = list(my_tuple)
converted_from_tuple[1] = 999
print(converted_from_tuple)
```

```
['a', 999, 4.4]
```

In practice, you can use a list almost anywhere you can use a tuple, though there are some important exceptions. One that you can already appreciate is that a tuple can be used as a key to a dictionary, but a list cannot:

```
dict_with_sequence_keys = {my_tuple : "Access this value using a tuple!"}
```

```
dict_with_sequence_keys[converted_from_tuple] = "This will not work"
```

```
---------------------------------------------------------------------
TypeError                                   Traceback (most recent call last)
Input In [39], in <cell line: 1>()
----> 1 dict_with_sequence_keys[converted_from_tuple] = "This will not work"

TypeError: unhashable type: 'list'
```

Admittedly, the error that this produces is a bit cryptic, but it relates directly to the fact that a mutable object is considered a bit unreliable because elements within it can change without notice.

5.4   Everything in Python Is an Object

Our discussion so far might give off the impression that some data types in Python are basic or special in some way. It is natural to think, for example, that strings, integers, and booleans are *primitive* data types, i.e., that they are built into the core of the language, behave in special ways, and can not be duplicated or modified. And this is true in many other programming languages. For example, in Java, there are exactly eight primitive data types. If you get bored of them, you are out of luck. You cannot just create new ones, i.e., a new type of string that behaves just like the primitive strings, but adds some additional functionality you think would be kind of cool to have.

Python is different: it does not *really* have any primitive data types. Python is a deeply *object-oriented* programming language, and in Python, *everything is an object*. Strings are objects, integers are objects, Booleans are objects. So are lists. So are dictionaries. *Everything* is an object in Python. We will spend more time talking about what objects are, and the deeper implications of everything being an object, at the end of this chapter. For now, let's focus on some of the practical implications for the way we write code.

5.4.1   *The Dot Notation*

Let's start with the dot (.) notation we use to indicate that we are accessing data or functionality inside an object. You have probably already noticed that there are two kinds of constructions we have been using in our code to do things with variables. There is the functional syntax, where we pass an object as an argument to a function:

```
len([2, 4, 1, 9])
```

```
4
```

And then there is the object-oriented syntax that uses the dot notation, which we saw when looking at some of the functionality implemented in strings:

```
phrase = "aPpLeS ArE delICIous"

phrase.lower()
```

```
'apples are delicious'
```

If you have some experience in another object-oriented programming language, the dot syntax will be old hat to you. But if you have mostly worked in data-centric languages (e.g., R or MATLAB), you might find it puzzling at first.

What is happening in the previous example is that we are calling a function attached to this object (this is called a *method* of the object) `lower()` *on* the `phrase` string itself. You can think of the dot operator `.` as expressing a relationship of belonging, or roughly translating as "look inside of." So, when we write `phrase.lower()`, we are essentially saying, "call the `lower()` method that is contained inside of `phrase`." (We are being a bit sloppy here for the sake of simplicity, but that is the gist of it.)

Note that `lower()` works on strings, but unlike functions like `len()` and `round()`, it is not a built-in function in Python. We cannot just call `lower()` directly:

```
lower("TrY to LoWer ThIs!")
```

```
---------------------------------------------------------------
NameError                       Traceback (most recent call last)
Input In [42], in <cell line: 1>()
----> 1 lower("TrY to LoWer ThIs!")

NameError: name 'lower' is not defined
```

Instead, it needs to be called via an instance that contains this function, as we did previously with `phrase`.

Neither is `lower()` a method that is available on *all* objects. For example, this will not work:

```
num = 6

num.lower()
```

```
-------------------------------------------------------------------
AttributeError                          Traceback (most recent call last)
Input In [43], in <cell line: 3>()
      1 num = 6
----> 3 num.lower()

AttributeError: 'int' object has no attribute 'lower'
```

Integers, as it happens, do not have a method called `lower()`. And neither do most other types. But strings do. And what the `lower()` method does, when called from a string, is return a lowercase version of the string to which it is attached. But that functionality is a feature of the string type itself, and *not* of the Python language in general.

Later, we will see how we define new types (or classes), and specify what methods they have. For the moment, the main point to take away is that almost all functionality in Python is going to be accessed via objects. The dot notation is ubiquitous in Python, so you will need to get used to it quickly if you are used to a purely functional syntax.

INSPECTING OBJECTS

One implication of everything being an object in Python is that we can always find out exactly what data an object contains and what methods it implements by inspecting it in various ways.

We will not look very far under the hood of objects in this chapter, but it is worth knowing about a couple of ways of interrogating objects that can make your life easier.

First, you can always see the type of an object with the built-in `type()` function:

```
msg = "Hello World!"

type(msg)
```

```
str
```

Second, the built-in `dir()` function will show you all of the methods implemented on an object, as well as *static attributes*, which are variables stored within the object. Be warned that this will often be a long list, and some of the attribute names you see (mainly those that start and end with two underscores) will look a little wonky. We will talk about those briefly later.

```
dir(msg)
```

```
['__add__',
 '__class__',
 '__contains__',
 '__delattr__',
```

```
'__dir__',
'__doc__',
'__eq__',
'__format__',
'__ge__',
'__getattribute__',
'__getitem__',
'__getnewargs__',
'__gt__',
'__hash__',
'__init__',
'__init_subclass__',
'__iter__',
'__le__',
'__len__',
'__lt__',
'__mod__',
'__mul__',
'__ne__',
'__new__',
'__reduce__',
'__reduce_ex__',
'__repr__',
'__rmod__',
'__rmul__',
'__setattr__',
'__sizeof__',
'__str__',
'__subclasshook__',
'capitalize',
'casefold',
'center',
'count',
'encode',
'endswith',
'expandtabs',
'find',
'format',
'format_map',
'index',
'isalnum',
'isalpha',
'isascii',
'isdecimal',
'isdigit',
'isidentifier',
'islower',
'isnumeric',
'isprintable',
'isspace',
'istitle',
'isupper',
'join',
'ljust',
'lower',
'lstrip',
'maketrans',
'partition',
```

```
'replace',
'rfind',
'rindex',
'rjust',
'rpartition',
'rsplit',
'rstrip',
'split',
'splitlines',
'startswith',
'strip',
'swapcase',
'title',
'translate',
'upper',
'zfill']
```

That is a pretty long list! Any name in that list is available to you as an *attribute* of the object (e.g., my_var.values(), my_var.__class__), meaning that you can access it and call it (if it is a function) using the dot notation. Notice that the list contains all of the string methods we experimented with earlier (including lower), as well as many others.

Exercise

Find the methods associated with "int" objects. Are they different from the methods associated with "float" objects?

5.5   Control Flow

Like nearly every other programming language, Python has several core language constructs that allow us to control the flow of our code, i.e., the order in which functions get called and expressions are evaluated. The two most common ones are conditionals (if-then statements) and for-loops.

5.5.1   Conditionals

Conditional (or if-then) statements allow our code to branch, meaning we can execute different chunks of code depending on which of two or more conditions is met. For example:

```
mango = 0.2

if mango < 0.5:
    print("Mangoes are super cheap; get a bunch of them!")
elif mango < 1.0:
    print("Get one mango from the store.")
else:
    print("Meh. I don't really even like mangoes.")
```

```
Mangoes are super cheap; get a bunch of them!
```

The printed statement will vary depending on the value assigned to the `mango` variable. Try changing that value and see what happens when you rerun the code.

Notice that there are three statements in the previous code: `if`, `elif` (which in Python stands for "else if"), and `else`. Only the first of these (i.e., `if`) is strictly necessary; the `elif` and `else` statements are optional.

Exercise

There can arbitrarily be many `elif` statements. Try adding another one to the previous code that executes only in the case that mangos are more expensive than 2.0 and less expensive than 5.0.

5.5.2 Loops

For-loops allow us to iterate (or loop) over the elements of a collection (e.g., a list) and perform the same operation(s) on each one. As with most things Python, the syntax is quite straightforward and readable:

```
for elem in random_stuff:
    print(elem)
```

```
eleventy
apple
7.14
banana
88
```

Here we loop over the elements in the `random_stuff` list. In each iteration (i.e., for each element), we assign the value to the temporary variable `elem`, which only exists within the scope of the `for` statement (i.e., `elem` will not exist once the for-loop is done executing). We can then do whatever we like with `elem`. In this case, we just print its value.

LOOPING OVER A RANGE

While we can often loop directly over the elements in an array (as in the previous example), it is also very common to loop over a range of integer indices, which we can then use to access data stored in some sequential form in one or more collections. To facilitate this type of thing, we can use Python's built-in `range()` function, which produces a sequence of integers starting from zero and stopping before the passed value:

```
num_elems = len(random_stuff)

for i in range(num_elems):
    val = random_stuff[i]
    print(f"Element {i}: {val}")
```

```
Element 0: eleventy
Element 1: apple
Element 2: 7.14
Element 3: banana
Element 4: 88
```

Exercise

The content that was printed in each iteration of the loop in the last example is formatted using an *f-string*. This is a way to compose strings that change based on information from the code surrounding them. An f-string is a string that has the letter "f" before it, as in this example, and it can contain segments enclosed by curly braces ({ and }) that contain Python statements. In this case, the Python statements in each curly bracket are variable names, and the values of the variables at that point in the code are inserted into the string, but you could also insert small calculations that produce a result that then gets inserted into the string at that location. As an exercise, rewrite the previous code so that in each iteration through the loop the value of i and the value of i squared are both printed. Hint: powers of Python numbers are calculated using the ** operator.

5.5.3   Nested Control Flow

We can also nest conditionals and for-loops inside one another (as well as inside other compound statements). For example, we can loop over the elements of random_stuff, as shown previously, but keeping only the elements that meet some condition, e.g., only those elements that are strings:

```python
# create an empty list to hold the filtered values
strings_only = []

# loop over the random_stuff list
for elem in random_stuff:
    # if the current element is a string...
    if isinstance(elem, str):
        # ...then append the value to strings_only
        strings_only.append(elem)

print("Only the string values:", strings_only)
```

```
Only the string values: ['eleventy', 'apple', 'banana']
```

5.5.4 Comprehensions

In Python, for-loops can also be written in a more compact way known as a list comprehension (there are also dictionary comprehensions, but we will leave that for you to look up as an exercise). List comprehensions are just a bit of *syntactic sugar*, meaning they are just a different way of writing the same code but do not change the meaning in any way. Here is the list comprehension version of the for-loop we wrote previously:

```
p = [print(elem) for elem in random_stuff]
```

```
eleventy
apple
7.14
banana
88
```

We can also embed conditional statements inside list comprehensions. Here is a much more compact way of writing the string-filtering snippet we wrote previously:

```
strings_only = [elem for elem in random_stuff if isinstance(elem, str)]

print("Only the string values:", strings_only)
```

```
Only the string values: ['eleventy', 'apple', 'banana']
```

List comprehensions can save you quite a bit of typing once you get used to reading them, and you may eventually even find them clearer to read. It is also possible to nest list comprehensions (equivalent to for-loops within for-loops), though that power should be used sparingly, as nested list comprehensions can be difficult to understand.

Exercise

Using a comprehension, create a list where each element is a tuple. The first element in each tuple should be the index of the element in `random_stuff` and the second element of the tuple should be its square.

5.5.5   *Whitespace is Syntactically Significant*

One thing you might have noticed when reading the conditional statements and for-loops shown earlier is that we always seem to indent our code inside these statements. This is not a matter of choice; Python is a bit of an odd duck among programming languages, in that it imposes strong rules about how whitespace can be used (i.e., whitespace is *syntactically significant*). This can take a bit of getting used to, but once you do, it has important benefits: there is less variability in coding style across different Python programmers, and reading other people's code is often much easier than it is in languages without syntactically significant whitespace.

The main rule you need to be aware of is that whenever you enter a *compound statement* (which includes for-loops and conditionals, but also function and class definitions, as we will see subsequently), you have to increase the indentation of your code. When you exit the compound statement, you then decrease the indentation by the same amount.

The exact amount you indent each time is technically up to you. But it is strongly recommended that you use the same convention everyone else does (described in the Python style guide, known as PEP8[4]), which is to always indent or dedent by four spaces. Here is what this looks like in a block with multiple nested conditionals:

```python
num = 800

if num > 500:
    if num < 900:
        if num > 700:
            print("Great number.")
        else:
            print("Terrible number.")
```

```
Great number.
```

Exercise

Modify this same code so that you (1) consistently use a different amount of indentation (e.g., three spaces), and (2) break Python by using invalid indentation.

5.6   Namespaces and Imports

Python is a high-level, dynamic programming language, which people often associate with flexibility and lack of precision (e.g., as we have already seen, you do not have to declare the type of your variables when you initialize them in Python). But in some ways, Python is

4. https://peps.python.org/pep-0008/

much more of a stickler than most other dynamic languages about the way Python developers write their code. We just saw that Python is very serious about how you indent your code. Another thing that is characteristic of Python is that it takes *namespacing* very seriously.

If you are used to languages like R or MATLAB, you might expect to have hundreds of different functions available to call as soon as you fire up an interactive prompt. By contrast, in Python, the *built-in namespace*—the set of functions you can invoke when you start running Python—is very small.[5] This is by design: Python expects you to carefully manage the code you use, and it is particularly serious about making sure you maintain orderly namespaces.

In practice, this means that any time you want to use some functionality that is not built-in and immediately available, you need to explicitly *import* it from whatever module it is currently in, via an `import` statement. This can initially look strange, but once you get used to it, you will find that it substantially increases code clarity and almost completely eliminates naming conflicts and confusion about where some functionality came from.

5.6.1   Importing a Module

Conventionally, all import statements in a Python file are consolidated at the very top (though there are some situations where this is not possible). Here is what the most basic usage of `import` looks like:

```
import json
dummy = {'a': 1, 'b': 5}
json.dumps(dummy)
```

```
'{"a": 1, "b": 5}'
```

In this case, we begin by importing a module from Python's *standard library*, i.e., the set of libraries that come bundled with the Python interpreter in every standard installation. Once we import a module, we can invoke any of the functions located inside it, using the dot syntax you see in the previous example. In this case, we import the json module, which provides tools for converting to and from the JavaScript Object Notation (JSON) format. JSON is a widely used text-based data representation format. In the previous example, we take a Python dictionary and convert (or *dump*) it to a JSON string by calling the dumps() function.

Note that if we had not explicitly imported json, the json.dumps() call would have failed, because json would be undefined in our namespace. You can also try directly calling dumps() alone (without the json prefix) to verify that there is no such function available to you in Python's root namespace.

---

5. https://docs.python.org/3/library/functions.html

5.6.2   Importing From a Module

Importing a module by name gives us access to all of its internal attributes. But sometimes we only need to call a single function inside a module, and we might not want to have to type the full module name every time we use that function. In that case, we can import *from* the module:

```
# defaultdict is a dictionary that has default values for new keys.
from collections import defaultdict

# When initializing a defaultdict, we specify the default type of new values.
test_dict = defaultdict(int)

# this would fail with a normal dict, but with a defaultdict,
# a new key with the default value is created upon first access.
test_dict['made_up_key']
```

```
0
```

In this case, we import `defaultdict` directly *from* the `collections` module *into* our current namespace. This makes `defaultdict` available for our use. Note that `collections` itself is *not* available to us unless we explicitly import it (i.e., if we run `import collections`):

```
import collections
another_test_dict = collections.defaultdict(int)
```

5.6.3   Renaming Variables at Import Time

Sometimes the module or function we want to import has an unwieldy name. Python's import statements allow us to rename the variable we are importing on the fly using the `as` keyword:

```
from collections import defaultdict as dd

float_test_dict = dd(float)
```

For many commonly used packages, there are strong conventions about naming abbreviations, and you should make sure to respect these in your code. For example, it is standard to see `import numpy as np` and `import pandas as pd` (both are libraries that you will learn about in chapter 8 and chapter 9, respectively). Your code will still work fine if you use other variable names, but other programmers will have a slightly more difficult time understanding what you are doing. So be kind to others and respect the conventions.

5.7 Functions

Python would be of limited use to us if we could only run our code linearly from top to bottom. Fortunately, as in almost every other modern programming language, Python has *functions*: blocks of code that only run when explicitly called. Some of these are built into the language itself (or contained in the standard library's many modules we can import from, as we saw previously):

```
approx_pi = 3.141592

round(approx_pi, 2)
```

```
3.14
```

Here we use the `round()` function to round a float to the desired precision (two decimal places). The `round()` function happens to be one of the few dozen built-ins included in the root Python namespace out of the box, but we can easily define our own functions, which we can then call just like the built-in ones. Functions are defined like this:

```
def print_useless_message():
    print("This is a fairly useless message.")
```

Here, we are defining a new function called `print_useless_message`, which, as you might expect, can print a fairly useless message. Notice that nothing happens when we run this block of code. That is because all we have done is *define* the function; we have not yet *called* or *invoked* it. We can do that like this:

```
print_useless_message()
```

```
This is a fairly useless message.
```

5.7.1 Function Arguments and Return Values

Functions can accept *arguments* (or parameters) that alter their behavior. When we called `round()` in the previous example, we passed two arguments: the float we wanted to round, and the number of decimal places we wanted to round it to. The first argument is mandatory in the case of `round()`; if we try calling `round()` without any arguments (feel free to give it a shot), we will get an error. This should make intuitive sense to you because it would be pretty strange to try to round no value at all.

Functions can also explicitly *return* values to the user. To do this, we have to explicitly end our function with a `return` statement, followed by the variable(s) we want to return.

If a function does not explicitly end with a `return` statement, then the special value `None` we encountered earlier will be returned.

Let's illustrate the use of arguments by writing a small function that takes a single float as input, adds Gaussian noise (generated by the standard library's `random` module), and returns the result.

```python
import random

def add_noise(x, mu, sd):
    """Adds gaussian noise to the input.

    Parameters
    ----------
    x : number
        The number to add noise to.
    mu : float
        The mean of the gaussian noise distribution.
    sd : float
        The standard deviation of the noise distribution.

    Returns
    -------
    float
    """
    noise = random.normalvariate(mu, sd)
    return (x + noise)
```

The `add_noise()` function has three required parameters: The first (`x`) is the number we want to add noise to. The second (`mu`) is the mean of the Gaussian distribution to sample from. The third (`sd`) is the distribution's standard deviation.

Notice that we have documented the function's behavior inside the function definition itself using what is called a *docstring*.[6] This a good habit to get into, as good documentation is essential if you expect other people to be able to use the code you write (including yourself in the future). In this case, the docstring indicates to the user what the expected type of each argument is, what the argument means, and what the function returns. In case you are wondering why it is organized in just this way, that is because we are following the conventions of docstrings established by the NumPy project (and described in the NumPy docstring guide[7]).

Now that we have defined our noise-adding function, we can start calling it. Note that because we are sampling randomly from a distribution, we will get a different output every time we rerun the function, even if the inputs are the same.

```python
add_noise(4, 1, 2)
```

6. https://www.python.org/dev/peps/pep-0257/
7. https://numpydoc.readthedocs.io/en/latest/format.html

```
7.07164813088094
```

Exercise

Based on the function definition provided previously, define a new function that produces a sample of n numbers each of which is x with Gaussian noise of mean mu and standard deviation sd added to it. The return value should be a list of length n, itself a parameter to the function.

5.7.2   Function Arguments

Python functions can have two kinds of arguments: *positional* arguments, and *keyword* (or *named*) arguments.

POSITIONAL ARGUMENTS

Positional arguments, as their name suggests, are defined by position, and they *must* be passed when the function is called. The values passed inside the parentheses are mapped one-to-one onto the arguments, as we saw previously for add_noise(). That is, inside the add_noise() function, the first value is referenced by x, the second by mu, and so on.

If the caller fails to pass the right number of arguments (either too few or too many), an error will be generated:

```
add_noise(7)
```

```
--------------------------------------------------------------------------
TypeError                                 Traceback (most recent call last)
Input In [62], in <cell line: 1>()
----> 1 add_noise(7)

TypeError: add_noise() missing 2 required positional arguments: 'mu' and 'sd'
```

In this case, the call to the function fails because the function has three positional arguments, and we only passed one.

KEYWORD ARGUMENTS

Keyword arguments are arguments that are assigned a default value in the function *signature* (i.e., the top line of the function definition, that looks like def my_function(...)). Unlike positional arguments, keyword arguments are optional:

if the caller does not pass a value for the keyword argument, the corresponding variable will still be available inside the function, but it will have whatever value is defined as the default in the signature.

To see how this works, let's rewrite our `add_noise()` function so that the parameters of the Gaussian distribution are now optional:

```python
def add_noise_with_defaults(x, mu=0, sd=1):
    """Adds gaussian noise to the input.

    Parameters
    ----------
    x : number
        The number to add noise to.
    mu : float, optional
        The mean of the gaussian noise distribution.
        Default: 0
    sd : float, optional
        The standard deviation of the noise distribution.
        Default: 1

    Returns
    -------
    float
    """
    noise = random.normalvariate(mu, sd)
    return x + noise
```

This looks very similar, but we can now call the function without filling in `mu` or `sd`. If we do not pass in those values explicitly, the function will internally use the defaults (i.e., `0` in the case of `mu`, and `1` in the case of `sd`). Now, when we call this function with only one argument:

```python
add_noise_with_defaults(10)
```

```
8.159679194958816
```

Keyword arguments do not have to be filled in order, as long as we explicitly name them. For example, we can specify a value for `sd` but not for `mu`:

```python
# we specify x and sd, but not mu
add_noise_with_defaults(5, sd=100)
```

```
103.04682507683225
```

Note that if we did not specify the name of the argument (i.e., if we called `add_noise_with_defaults(5, 100)`, the function would still work, but the second value we pass would be interpreted as mu rather than sd because that is the order they were introduced in the function definition.

It is also worth noting that we can always explicitly name *any* of our arguments, including positional ones. This is extremely handy in cases where we are calling functions whose argument names we remember, but where we do not necessarily remember the exact order of the arguments. For example, suppose we remember that add_noise() takes the three arguments x, mu, and sd, but we do not remember if x comes before or after the distribution parameters. We can guarantee that we get the result we expect by explicitly specifying all the argument names:

```
add_noise(mu=1, sd=2, x=100)
```

```
101.69396484968273
```

To summarize, functions let us define a piece of code that can be called as needed and reused. We can define a default behavior and override it as necessary. There is a bit more to it, of course. For a few more details, you can go into more depth in the following, or you can skip forward to Section 5.8.

ARGUMENT UNPACKING WITH *ARGS AND **KWARGS

It sometimes happens that a function needs to be able to accept an unknown number of arguments. A very common scenario like this is where we have written a *wrapper* function that takes some input, does some operation that relies on only some of the arguments the user passed, and then hands off the rest of the arguments to a different function.

For example, suppose we want to write an arg_printer() function that we can use to produce a standardized display of the positional and keyword arguments used when calling some other *arbitrary* function. Python handles this scenario elegantly via special *args and **kwargs syntax in function signatures, also known as *argument unpacking*.

The best way to understand what *args and **kwargs do is to see them in action. Here is an example:

```
def arg_printer(func, *args, **kwargs):
    """
    A wrapper that takes any other function plus arguments to pass
    to that function. The arguments are printed out before the
    function is called and the result returned.

    Parameters
    ----------
    func : callable
        The function to call with the passed args.
    args : list, optional
```

```
        List of arguments to pass into func.
    kwargs : dict, optional
        Dict of keyword arguments to pass into func.

    Returns
    -------
    The result of func() when called with the passed arguments.
    """
    print("Calling function:", func.__name__)
    print("Positional arguments:", args)
    print("Keyword arguments:", kwargs)
    return func(*args, **kwargs)
```

This may seem a bit mysterious, and there are parts we will not explain right now (e.g., func.__name__). But try experimenting a bit with calling this function, and things may start to click. Here is an example to get you rolling:

```
arg_printer(add_noise, 17, mu=0, sd=5)
```

```
Calling function: add_noise
Positional arguments: (17,)
Keyword arguments: {'mu': 0, 'sd': 5}
```

```
18.91215070453928
```

What is happening here is that the first argument to arg_printer() is the add_noise() function we defined earlier. Remember, everything in Python is secretly an object, even functions! You can pass functions as arguments to other functions too; it is no different than passing a string or a list. A key point to note, however, is that what we are passing is, in a sense, the *definition* of the function. Notice how we did not add parentheses to add_noise when we passed it to arg_printer(). That is because we do not want to call add_noise yet; we are leaving it to the arg_printer() function to do that internally.

All the other arguments to arg_printer() after the first one are arguments that we actually want to pass to the add_noise() function when it is called internally by arg_printer(). The first thing arg_printer() does is print out the name of the function we just gave it, as well as all of the positional and keyword arguments we passed in. Once it has done that, it calls the function we passed in (add_noise()) and passes along all the arguments.

If the previous example does not make sense, do not worry! As we mentioned before, we are moving quickly, and the concepts from here on out start to get quite a bit denser. A good way to explore these ideas a bit better is to write your own code and experiment with things until they start to make sense.

Exercise

Replace `add_noise` with built-in functions like `min`, `len`, or `list`. What other arguments do you need to change?

5.8 Classes

The material we have covered so far in this chapter provides a brief overview of the most essential concepts in Python and should be sufficient to get you started reading and writing code in the language. If you are impatient to start working with scientific computing libraries and playing with neuroimaging data, you could stop here and move on to the next chapter. That said, there are some concepts we have not talked about yet that are quite central to understanding how Python really works under the hood, and you will probably get more out of this book if you understand them well. However, these things are quite a bit more difficult conceptually.

In particular, the object-oriented programming (OOP) paradigm, and Python's internal data model (which revolves around something called *magic methods*), are not very intuitive, and it can take some work to wrap your head around them if you have not seen OOP before. We will do our best to introduce these concepts here, but do not be surprised or alarmed if they do not immediately make sense to you—that is completely normal! You will probably find that much of this material starts to click as you work your way through the various examples in this book and start to write your own Python analysis code.

We have said several times now that everything in Python is actually an object. We are now in a position to unpack that statement. What does it mean to say that something is an object? How do objects get defined? And how do we specify what an object should do when a certain operation is applied to it? To answer these questions, we need to introduce the notion of *classes* and, more generally, the *object-oriented programming* paradigm.

5.8.1 What Is a Class?

A class is, in a sense, a kind of template for an object. You can think of it as a specification or a set of instructions that determine what an object of a given kind can do. It is very close in meaning to what we have already been referring to as the *type* of an object. There *is* technically a difference between types and classes in Python but it is quite subtle, and in day-to-day usage you can use the terms interchangeably and nobody is going to yell at you.

5.8.2 Defining Classes

So a class is a kind of template; okay, what does it look like? Well, minimally, it looks like very little. Here is a fully functional class definition:

```
class Circle:
    pass
```

That is it! We have defined a new Python class. In case you are wondering, the `pass` statement does nothing, it is used as a placeholder that tells the Python interpreter it should not expect any more code to follow.

5.8.3   Creating Instances

You might think that this empty `Circle` class definition is not very interesting because we obviously cannot *do* anything with it. But that is not entirely true. We can already *instantiate* this class if we like, which is to say we can create new objects whose behavior is defined by the `Circle` class. A good way to think about it is in terms of an "X is a particular Y" relationship. If we were to draw three different circles on a piece of paper, we could say that each one is an *instance* of a circle, but none of them would be the actual *definition* of a circle. That should give you an intuitive sense of the distinction between a class definition and instances of that class.

The syntax for creating an instance in Python is simple:

```
my_circle = Circle()
```

That is it again! We now have a new `my_circle` variable on our hands, which is an instance of class `Circle`.

If you do not believe that, it is easy to prove:

```
type(my_circle)
```

```
__main__.Circle
```

In case you are wondering, the reason that this appears to be a `__main__.Circle` object is that while we are running the code in this notebook, it is defined within a namespace (yes, namespaces again) called `__main__`, and the `type` function recognizes this fact. You can learn more about this particular namespace in the Python documentation.[8]

A NOTE ON NOMENCLATURE

You may have already noticed the naming convention we have used throughout this tutorial: our variable names are always composed of lower-case characters, with words separated by underscores. This is called *snake_case*. You will also note that class names are

---

8. https://docs.python.org/3/library/__main__.html

capitalized (technically, they are in *CamelCase*). Both of these are standard conventions in the Python community, and you should get in the habit of following both.

5.8.4 Making It Do Things

The `Circle` class definition we wrote in the previous example was perfectly valid, but not terribly useful. It did not define any new behavior, so any instances of the class we created would not do anything more than base objects in Python can do (which is not very much).

Let's fix that by filling in the class a bit.

```python
from math import pi

class Circle:

    def __init__(self, radius):
        self.radius = radius

    def area(self):
        return pi * self.radius**2
```

There is not much code to see here, but conceptually, a lot is going on. Let's walk through this piece by piece.

We start by importing the variable `pi` from the standard library `math` module.

Then, we start defining the class. First, observe that we have defined what looks like two new functions inside the class. Technically, these are *methods* and not functions, because they are *bound* to a particular object. But the principle is the same: they are chunks of code that take arguments, do some stuff, and then (possibly) return something to the caller.

You will also note that both methods have `self` as their first argument. This is a requirement: all instance methods have to take a reference to the current instance (conventionally named `self`) as their first argument (there are also *class* methods and *static* methods, which behave differently, but we will not cover those). This reference is what will allow us to easily *maintain state* (that is, to store information that can vary over time) inside the instance.

Now let's walk through each of the defined methods. First, there is `__init__()`. The leading and trailing double underscores indicate that this is a special kind of method called a *magic* method; we will talk about those a bit more later. For the moment, we just have to know that `__init__()` is a special method that gets called whenever we create a new instance of the class. So, when we write a line like `my_circle = Circle()`, what happens under the hood is that the `__init__()` method of `Circle` gets executed.

In this case, observe that `__init__()` takes a single argument (other than `self`, that is): a `radius` argument. And further, the only thing that happens inside `__init__()` is that we store the value of the input argument `radius` in an *instance attribute* called radius. We do this by assigning to self.radius. Remember: self is a reference to the current instance, so the newly created instance that is returned by Circle() will have that radius value set in the .radius attribute.

This code should make this a bit clearer:

```
my_circle = Circle(4)
print(my_circle.radius)
```

```
4
```

Next, let's look at the area() method. This one takes no arguments (again, self is passed automatically; we do not need to, and should not, pass it ourselves). That means we can just call it and see what happens. Let's do that:

```
my_circle.area()
```

```
50.26548245743669
```

When we call area(), what we get back is the area of our circle based on the radius stored in the instance at that moment. Note that this area is only computed when we call area(), and is not computed in advance. This means that if the circle's radius changes, so too will the result of area():

```
my_circle.radius = 9
print(my_circle.area())
```

```
254.46900494077323
```

Exercise

1. Add to the implementation of Circle another method that calculates the circumference of the circle.
2. Implement a class called Square that has a single attribute .side, and two methods for area and circumference.

5.8.5   *Magic Methods*

There is a *lot* more we could say about how classes work in Python, and about object-oriented programming in general, but this is just a brief introduction, so we have to be picky. Let's introduce just one other big concept: *magic methods*. The concepts we will discuss in this last section are fairly advanced, and are not usually discussed in introductions to Python, so as we have said several times now, do not worry if they do not immediately click for you. We promise you will still be able to get through the rest of the book just fine! The reason we decided to cover magic methods here is that, once you *do* understand them well, you will have a substantially deeper grasp on how Python works, and you will probably see patterns and make connections that you otherwise would not.

Magic methods of objects, as we have seen a couple of times now, start and end with a double underscore: __init__, __getattr__, __new__, and so on. As their names suggest, these methods are magic, at least in the sense that they appear to add some magic to a class's behavior. We have already talked about __init__, which is a magic method that gets called any time we create a new instance of a class. But there are many others.

The key to understanding how magic methods work is to recognize that they are usually called implicitly when a certain operation is applied to an object, even if it does not look like the magic method and the operation being applied have anything to do with each other; that is what makes them magic!

Remember how we said earlier that everything in Python is an object? Well, now we are going to explore one of the deeper implications of that observation, which is that *all operators in Python are just cleverly disguised method calls*. That means that when we write even an expression as seemingly basic as 4 * 3 in Python, it is implicitly converted to a call to a magic method on the first operand (4), with the second operand (3) being passed in as an argument.

This is a bit hard to explain abstractly, so let's dive into an example. Start with this naive arithmetic operation:

```
4 * 3
```

```
12
```

No surprises there. But here is an equivalent way to write the same line, which makes clearer what is happening under the hood when we multiply one number by another:

```
# 4 is a number, so we have to wrap it in parentheses to prevent a syntax error.
# but we wouldn't have to do this for other types of variables (e.g., strings).
(4).__mul__(3)
```

```
12
```

Remember the dot notation? Here, __mul__ is a (magic) method implemented in the integer class. When Python evaluates the expression 4 * 3, it calls __mul__ on the first integer, and hands it the second one as an argument. See, we were not messing around when we said *everything is an object in Python*. Even something as seemingly basic as the multiplication operator is an alias to a method called on an integer object.

5.8.6   The Semantics of *

Once we recognize that Python's * operator is just an alias to the __mul__ magic method, we might start to wonder if this is *always* true. Does every valid occurrence of * in Python code imply that the object just before the * must be an instance of a class that implements the __mul__ method? The answer is yes! The result of an expression that includes the * operator (and for that matter, every other operator in Python, including things like == and &) is entirely dependent on the receiver object's implementation of __mul__.

Just to make it clear how far-reaching the implications of this principle are, let's look at how a couple of other built-in Python types deal with the * operator. Let's start with strings. What do you think will happen when we multiply a string object by 2?

```
"apple" * 2
```

```
'appleapple'
```

There is a good chance this was *not* the behavior you expected. Many people intuitively expect an expression like "apple" * 2 to produce an error because we do not normally think of strings as a kind of thing that can be multiplied. But remember: in Python, the multiplication operator is just an alias for a __mul__ call. And there is no particular reason a string class *should not* implement the __mul__ method; why not define *some* behavior for it, even if it is counterintuitive? That way users have a super easy way to repeat strings if that is what they want to do.

What about a list?

```
random_stuff * 3
```

```
['eleventy',
 'apple',
 7.14,
 'banana',
 88,
 'eleventy',
 'apple',
 7.14,
 'banana',
 88,
 'eleventy',
 'apple',
 7.14,
 'banana',
 88]
```

List multiplication behaves a lot like string multiplication: the result is that the list is repeated *n* times.

What about dictionary multiplication?

```
fruit_prices * 2
```

```
-----------------------------------------------------------------
TypeError                         Traceback (most recent call last)
Input In [80], in <cell line: 1>()
----> 1 fruit_prices * 2

TypeError: unsupported operand type(s) for *: 'dict' and 'int'
```

Finally, we encounter an outright failure! It appears Python dictionaries cannot be multiplied. This presumably means that the dict class does not implement __mul__ (you can verify this for yourself by inspecting any dictionary using dir()).

5.8.7 Other Magic Methods

Most of the magic methods in Python do something very much like what we saw for the multiplication operator. Consider the following operators: +, &, /, %, and <. These map, respectively, onto the magic methods __add__, __and__, __truediv__, __mod__, and __lt__. Many others follow the same pattern.

There are also magic methods that are tied to built-in functions rather than operators (e.g., when you call len(obj), that is equivalent to calling obj.__len__), or that are triggered by certain events (e.g., __getattr__ is called when a requested attribute is not found in an object).

In practice, you will not need to know much about magic methods unless you start writing a lot of classes of your own; if you are interested you can find a full description of all the magic methods in the official Python documention.[9] We spent a lot of time talking about them mainly because they are a good way to convey some deep insights about the data model at the core of the Python language.

5.8.8 Hungry Circles

Let's come full circle now (awful pun intended) and revisit the Circle class we defined earlier. The last thing we will do in this chapter is to add a magic method to our Circle class. This will nicely tie together a lot of different threads we have covered.

What we are going to do is give instances of class Circle the ability to eat other circles. When given Python code like this:

9. https://docs.python.org/3/reference/datamodel.html

```
c1 = Circle(4)
c2 = Circle(2)
c1 * c2
```

we want the first circle to grow its radius by exactly the amount required for its new area to equal the sum of the two circles' previous areas. Here is our updated implementation:

```
from math import pi, sqrt

class Circle:

    def __init__(self, radius):
        self.radius = radius

    def __mul__(self, prey):
        new_area = self.area() + prey.area()
        self.radius = sqrt(new_area / pi)

    def area(self):
        return pi * self.radius**2
```

The only change here is the addition of the __mul__ method.
Let's see if the previous example did what we wanted:

```
c1 = Circle(4)
c2 = Circle(2)

# Now the important part: c1 eats c2!
c1 * c2
```

Well, we did not get an error, so that is a good sign. Let's inspect c1 and see if it has been updated as we expect. Remember, we expect c1 to have eaten c2, which means its radius should grow, and its area should be the sum of both previous areas.

```
print("Radius of c1 after gorging on c2:", c1.radius)
print("Area of c1 after gorging on c2:", c1.area())
```

```
Radius of c1 after gorging on c2: 4.47213595499958
Area of c1 after gorging on c2: 62.83185307179588
```

It worked!

The only slightly dissatisfying feature of our implementation is that, after c1 eats c2 and expands itself accordingly, c2 is somehow still around to tell the tale. This probably violates some physical conservation law, but we will overlook that here. For reasons we will not get into, it is not trivial to delete c2 from inside c1. (There are good reasons for this, and the fact that we cannot easily make some of our circles wink out of existence from inside the belly of other circles might lead us to suspect we have architected our code suboptimally. But that is a problem for a different book.)

Exercise

Add a __mul__ method to your implementation of the Square class that follows the same principles as the __mul__ method of the Circle class, changing both the side attribute of the calling object, as well as the return value of area as it swallows the other object.

5.9 Additional Resources

This chapter provided a high-level look at some of the main features of the Python language—some basic, some more advanced. To develop a stronger working familiarity with the language, you will need to roll up your sleeves and start writing some code. One of the best ways to learn is to pick a small problem that interests or matters to you in some way (e.g., parsing some text data you have lying around), and search the web for help every time you run into problems. There is no shame in consulting the internet! All programmers do it.

If you prefer to have more structure than that, there are hundreds of excellent, and mostly free, resources online to help you on your way. A couple of good ones include:

- A Whirlwind Tour of Python[10] is an excellent intro to Python by Jake VanderPlas; Jupyter notebooks are available in the book GitHub repo.[11]
- Allen Downey's "Think Python"[12] is another excellent introduction to the language.

---

10. http://www.oreilly.com/programming/free/files/a-whirlwind-tour-of-python.pdf
11. https://github.com/jakevdp/WhirlwindTourOfPython
12. http://greenteapress.com/wp/think-python-2e/

6

The Python Environment

The previous chapter introduced the core elements of the Python language. If you read carefully, interacted with the examples, and did some of the exercises, you should now have a basic understanding of what Python code looks like and be in a good position to start writing your own programs. You might already be itching to put this knowledge into practice and start working with neuroimaging data. Do not worry; we will get there! But first, we are going to spend some time talking about two underappreciated topics that will make you a more effective data scientist. In this section, we will discuss best practices for writing code on your own, that is, practices that will allow you to do data science in Python more comfortably or efficiently. We will cover choosing a good code editor and some basic methods for debugging, testing, and profiling your code.

In the next chapter, we will talk about the best practices for sharing and maintaining Python projects, namely a manner of operating that will make it easier for you to share your code with others, and will increase the likelihood of other people using your code and reciprocally contributing to it.

We freely admit that most readers will not find the material we are going to cover in these next two chapters super exciting. That is okay! It is understandable if you are more excited about learning to fit fancy machine learning models to brain imaging data than about learning to automatically test your code. We are too! But we strongly encourage you to take these sections just as seriously as the rest of the book. We have consistently observed that people who do so tend to progress more rapidly. We are confident that you will be a considerably more efficient programmer and data scientist if you learn just a little bit about tooling and best practices at the front end. In our experience, much of this material has an *I wish I had known about this sooner* flavor; i.e., people tend to learn about it fairly late into their development as data scientists or programmers, and often kick themselves for not taking the time to learn it sooner. Our goal is to try to spare you this type of reaction.

6.1 Choosing a Good Editor

The first order of business for every data scientist should be to choose and set up a good *development environment*. When we say development environment we do not mean the physical environment in which you work (although that is important too), but rather the programs that you should install on your computer and set up to provide you with a smooth path from ideas to software implementation. Python code is plain text, so to write it you are going to need an application that facilitates the editing of plain text; i.e., a text editor of some kind. This seems straightforward, and in principle you can write great code in *any* text editor, including simple editors with very few bells and whistles like Notepad or TextEdit that come packaged with the Windows or OS X operating systems. But it is well worth your time to pick an editor that is specifically designed with code development—and ideally, Python code development—in mind. There is a little bit of a learning curve involved in getting used to these programs and being able to take advantage of all of their features, but once you get used to what a good editor provides, we are confident you will not dream of going back.

Here are just a few of the things a good editor will do for you that a generic text editor like Notepad will not support:

- Syntax and error highlighting: the editor uses different colors and styling to display different elements of your code and identify errors.
- Automatic code completion: as you type variable or function names, the editor shows you available valid completions.
- Code formatting: code is automatically formatted to meet the language's styling rules or conventions.
- Integrated code and test execution: you can execute a piece of code, or tests, directly from inside the editor.
- Built-in debugging tools: you can step through your code and visualize your workspace while your program runs.

We will not cover all of these here, but to give you a sense of just some of the benefits, in Fig. 6.1, we show the same snippet of Python code, displayed in a text editor not customized for editing code (TextEdit) and in one of the most popular editors for software development (VS Code). Hopefully, you can immediately appreciate how much more pleasant it is to write code on the right side than on the left. And this is just scratching the surface of what a good editor offers.

6.1.1 Evaluating the Options

There are a *lot* of editors out there. How do we choose between them? One helpful heuristic is to look at what is popular. Some of the most widely used Python editors are (in no particular order) VS Code, PyCharm, Spyder, Sublime Text, Atom, Vim, and emacs. There are some major differences between these. Most notably, Vim and emacs are highly configurable Unix-based text editors that historically do not have a graphical user interface (GUI), whereas the others all run as windowed cross-platform applications. In terms of feature sets, there is relatively little any of them can do that the others cannot; in part, this is because most of them have a fairly well-developed plug-in that allows you to easily

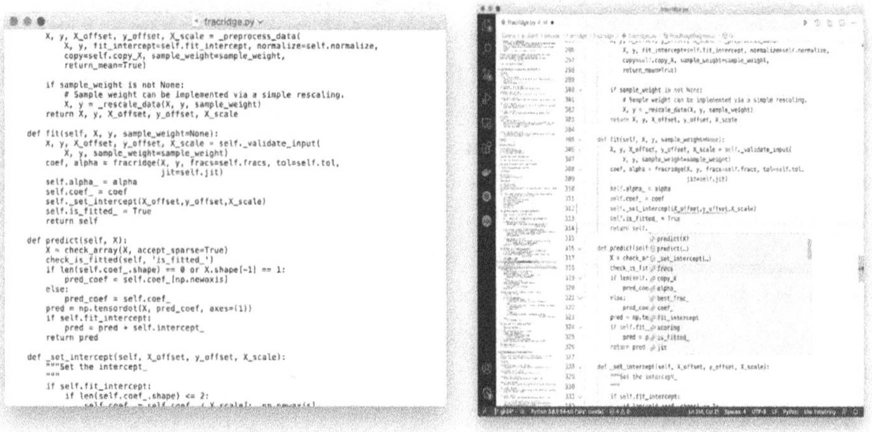

FIGURE 6.1. Python code in TextEdit (left) and VS Code (right).

extend the functionality of the editor by installing extensions from a central (and often very large) library.

If you can spare the time, we suggest trying out a few of these editors and seeing what works best for you. If you do not have the time, our recommendation would be to use VS Code. It has an extremely rich set of features, is easily extended via an unparalleled plug-in ecosystem, and has excellent developer support (it is built by Microsoft, though the product is free and open-source).

6.2   Debugging

When programmers talk about *writing* code, what they often euphemistically mean is *debugging* code. Almost every programmer spends a good chunk of their time trying to figure out why the code they just wrote is generating errors (or *raising exceptions*, in Python parlance). No matter how good a programmer you are (or become), you are not going to be able to avoid debugging. So it is in your best interest to learn to use a few tools that speed up the process. Here we will cover just a couple of tools that are readily available to you in Python's standard library and should get you on your way.

Let's start with some code that does not work exactly right. Consider the following function:

```python
def add_last_elements(list1, list2):
    """Returns the sum of the last elements in two lists."""
    return list1[-1] + list2[-1]
```

This function will work fine for many inputs, but if we try to call it like this, it will fail:

```python
add_last_elements([3, 2, 8], [])
```

```
---------------------------------------------------------------
IndexError                              Traceback (most recent call last)
Input In [3], in <cell line: 1>()
----> 1 add_last_elements([3, 2, 8], [])

Input In [2], in add_last_elements(list1, list2)
      1 def add_last_elements(list1, list2):
      2     """Returns the sum of the last elements in two lists."""
----> 3     return list1[-1] + list2[-1]

IndexError: list index out of range
```

The `IndexError` exception produced here (an *exception* is Python's name for an error) is a pretty common one. What that error message (`list index out of range`) means is that the index we are trying to access within an iterable object (in our case, a list) is outside of the valid bounds—for example, we might be trying to access the fourth element in a list that only contains two elements.

In this particular case, the problem is that our function is trying to sum the elements at the `-1` index in each list (i.e., the last element), but one of the lists we passed in is empty, and hence does not have a last element. The result is that the Python interpreter aborts execution and complains loudly that it cannot do what we are asking it to do.

Once we understand the problem, it is easy to fix. We could either change our input, or handle this edge case explicitly inside our function (e.g., by ignoring empty lists in the summation). But let's suppose we had no immediate insight into what was going wrong. How would go about debugging our code? We will explore a few different approaches.

6.2.1 Debugging with Google

A first, and generally excellent, way to debug Python code is to use a search engine. Take the salient part of the exception message (usually the very last piece of the output)—in this case, `IndexError: list index out of range`—and drop it into Google. There is a good chance you will find an explanation or solution to your problem on Stack Overflow or some other forum in fairly short order.

One thing we have learned from many years of teaching people to program is that beginners often naively assume that experienced software developers have successfully internalized everything there is to know about programming, and that having to look something up on the internet is a sign of failure or incompetence. Nothing could be farther from the truth. While we do not encourage you to just blindly cut and paste solutions from Stack Overflow into your code without at least *trying* to understand what is going on, attempting to solve every problem yourself from first principles would be a horrendously inefficient way to write code, and it is important to recognize this. One of our favorite teaching moments occurred when one of us ran into a problem during a live demo and started Googling in real time to try and diagnose the problem. Shortly afterward, one of the members of the audience tweeted that they had not realized that even experienced developers have to rely on Google to solve problems. That brought home to us how deep this misconception runs for many beginners; we are mentioning it here so that you can nip it in the bud. There is nothing wrong with consulting the internet when you run into problems! In fact, you would be crazy *not* to.

6.2.2 Debugging with Print and Assert

Not every code problem can be solved with a few minutes of Googling. And even when it can, it is often more efficient to just try to figure things out ourselves. The simplest way to do that is to liberally sprinkle `print()` functions or `assert` statements throughout our code.

Let's come back to the exception we generated in the previous example. Since we passed the list arguments to `add_last_elements()` explicitly, it is easy to see that the second list is empty. But suppose those variables were being passed to our function from some other part of our codebase. Then we might not be able to just glance at the code and observe that the second list is empty. We might still hypothesize that the culprit is an empty list, but we need to do a bit more work to confirm that hunch. We can do that by inserting some `print()` or `assert` calls in our code. In this case, we will do both:

```python
def add_last_elements(list1, list2):
    """Returns the sum of the last elements in two lists."""

    # Print the length of each list
    print("list1 length:", len(list1))
    print("list2 length:", len(list2))

    # Add an assertion that requires both lists to be non-empty
    assert len(list1)>0 and len(list2)>0, "At least one of the lists is empty"

    return list1[-1] + list2[-1]

# We'll define these variables here, but imagine that they were
# passed from somewhere else, so that we couldn't discern their
# contents just by reading the code!
list1 = [3, 2, 8]
list2 = []

add_last_elements(list1, list2)
```

```
list1 length: 3
list2 length: 0
```

```
---------------------------------------------------------------------------
AssertionError                            Traceback (most recent call last)
Input In [4], in <cell line: 19>()
     16 list1 = [3, 2, 8]
     17 list2 = []
---> 19 add_last_elements(list1, list2)

Input In [4], in add_last_elements(list1, list2)
      6 print("list2 length:", len(list2))
      8 # Add an assertion that requires both lists to be non-empty
----> 9 assert len(list1)>0 and len(list2)>0, "At least one of the lists is empty"
     11 return list1[-1] + list2[-1]

AssertionError: At least one of the lists is empty
```

Here, our `add_last_elements()` invocation still fails, but there are a couple of differences. First, before the failure, we get to see the output of our `print` calls, which shows us that `list2` has `zero` elements.

Second, we now get a different exception! Instead of an `IndexError`, we now get an `AssertionError`, with a custom message we wrote. What is going on here? At first glance, the line contain our assertion might seem a bit cryptic. Why does our code fail when we write `assert len(list1)>0 and len(list2)>0`?

The answer is that the `assert` statement's job is to evaluate the expression that immediately follows it (in this case, `len(list1)>0 and len(list2)>0`). If the expression evaluates to `True`, the interpreter continues happily on to the next line of code. But if it evaluates to `False`, an exception will be triggered, providing feedback to the programmer about where things are going wrong. In this case, the expression evaluates to `False`. So our assertion serves to ensure that both of our lists contain at least one value. If they do not, we get an `AssertionError`, together with a helpful error message.

6.2.3 Debugging with *pdb*

You can go a pretty long way with just `print()` and `assert`. One of us (we are not going to divulge who) debugs almost exclusively this way. That said, one big limitation of this approach is that we often end up rerunning our code many times, each time with different `print()` calls or assertions, just to test different hypotheses about what is going wrong. In such cases, it is often more efficient to freeze execution at a particular moment in time and step into our code. This allows us to examine all the objects in our workspace as the Python interpreter is seeing them at that moment. In Python, we can do this using the *Python debugger*, available in the standard library's `pdb` module.

There are different ways to interact with code through `pdb`, and here we will highlight just one: `set_trace()`. When you insert a `set_trace()` call into your code, the Python interpreter will halt when it comes to that line, and will drop you into a command prompt that lets you interact (in Python code) with all of the objects that exist in your workspace at that moment in time. Once you are done, you use the `continue` command to keep running. Here is an example:

```python
import pdb

def add_last_elements(list1, list2):
    """Returns the sum of the last elements in two lists."""
    pdb.set_trace()
    return list1[-1] + list2[-1]
```

Now when we call `add_last_elements()`, we get dropped into a debugging session, and we can directly inspect the values of `list1` and `list2` (or run any other code we like). For example, when running this code in Jupyter, we would be dropped into a new environment with an `ipdb>` command prompt (`ipdb` is the Jupyter version of `pdb`). We can type commands at this prompt and see the terminal output on the next line. There are different commands that you can run in the debugger (try executing `help` or `h` to get a full list of these), but one of the most common is the `p` or `print` command, which will print the value of some variable into the debugger output. For example, we can check the length of each input list by `p len(list2)` and `p len(list1)`, to observe that `list2` is empty (i.e., it has length 0). We can also reproduce the original exception by explicitly trying to access the last element: `list2[-1]` and observing the same failure. We can even replace `list2` with a new (nonempty) list before we continue the original program by executing `c` at the debugger command prompt. The fact that our function successfully returns a value once we replace `list2` confirms that we have correctly identified (and can fix) the problem.

6.3   Testing

Almost everyone who writes any amount of code will happily endorse the assertion that it is a good idea to test one's code. But people have very different ideas about what testing code entails. In our experience, what scientists who write research-oriented code mean when they tell you they have *tested* their code is quite different from what professional software developers mean by it. Many scientists think that they have done an adequate job testing a script if it runs from start to finish without crashing, and the output more or less matches one's expectations. This kind of haphazard, subjective approach to testing is certainly better than doing *no* testing, but it would horrify many software developers nevertheless.

Our goal in this section is to convince you that there is a lot of merit to thinking at least a little bit like a professional developer. We will walk through a few approaches you can use to make sure your code is doing what it is supposed to. The emphasis in all cases is on *automated* and *repeated* testing. Rather than relying on our subjective impression of whether or not our code works, what we want to do is write more code that tests our code. We want the test code to be so easy to run that we will be able to run it every time we make any changes to the code. That way we dramatically reduce the odds of finding ourselves thinking, *hmmm, I do not understand why this code is breaking—my labmate says it ran perfectly for them last month!*

6.3.1   Writing Test Functions

A central tenet of software testing can be succinctly stated like this: *code should be tested by code.* Most code is written in a modular way, with each piece of code (e.g., a function) taking zero or more inputs, doing some well-defined task, and returning zero or more outputs. This means that, so long as you know what inputs and/or outputs that code is expecting, you should be able to write a second piece of code that checks to make sure that the first piece of code produces the expected output when given valid input. This is sometimes also called *unit testing*, because we are testing our code one atomic unit at a time.

The idea is probably best conveyed by example. Let's return to the add_last_elements function we wrote previously. We will adjust it a bit to handle the empty list case that caused us the problems in the previous example:

```python
def add_last_elements(list1, list2):
    """Returns the sum of the last elements in two lists."""
    total = 0
    if len(list1) > 0:
        total += list1[-1]
    if len(list2) > 0:
        total += list2[-1]
    return total
```

Now let's write a test function for add_last_elements. We do not get points for creativity when naming our tests; the overriding goal should be to provide a clear description. Also, for reasons that will become clear in chapter 7, Python test function names conventionally start with test_. So we will call our test function test_add_last_elements. Here it is:

```python
def test_add_last_elements():
    """Test add_last_elements."""
    # Last elements are integers
    assert add_last_elements([4, 2, 1], [1, 2, 3, 999]) == 1000
    # Last elements are floats
    assert add_last_elements([4.833], [0.8, 4, 2.0]) == 6.833
    # One list is empty
    assert add_last_elements([], [3, 5]) == 5
```

Notice that our test contains multiple assert statements. Each one calls add_last_elements with different inputs. The idea is to verify that the function behaves as we would expect it to not just when we pass it ideal inputs, but also when we pass it *any* input that we deem valid. In this case, for example, our first assertion ensures that the function can handle integer inputs, and the second assertion ensures it can handle floats. The third assertion passes one empty list, which allows us to verify that this case is now properly handled.

Let's see what happens when we run our new test function:

```python
test_add_last_elements()
```

The answer is. . . nothing! Nothing happened. That is a good thing in this case because it means that every time the Python interpreter reached a line with an assertion, the logical statement on that line evaluated to True and the interpreter had no reason to raise an AssertionError. In other words, all of the assertions passed (if we wanted an explicit acknowledgment that everything is working, we could always print() a comforting message at the end of the test function). If you doubt that, you are welcome to change one of the test values (e.g., change 5 to 4) and observe that you now get an exception when you rerun the test.

Having a test function we can use to make sure that `add_last_elements` works as expected is incredibly useful. It makes our development and debugging processes far more efficient and much less fragile. When we test code manually, we treat it monolithically: if one particular part is breaking, that can be very hard to diagnose. We also get much better coverage: when we do not have to rerun code manually hundreds of times, we can afford to test a much wider range of cases. And of course, the tests are systematic and (usually) deterministic: we know that the same conditions will be evaluated every single time.

The benefits become even clearer once we have multiple test functions, each one testing a different piece of our codebase. If we run all of our tests in sequence, we can effectively determine not only *whether* our code works as we expect it to, but also, in the event of failure, *how* it is breaking. When you test your code holistically by running it yourself and seeing what happens, you often have to spend a lot of time trying to decipher each error message by working backward through the entire codebase. By contrast, when you have a bunch of well-encapsulated unit test functions, you will typically observe failures in only some of the tests (often only in a single one). This allows you to very quickly identify the source of your problem.

6.4   Profiling Code

One of the things that is enabled by testing is the process of code *refactoring*. This is a process whereby software is improved without fundamentally changing its interface. A common example of refactoring is introducing improvements to the performance of the code, i.e., accelerating the runtime of a piece of code but leaving the functionality of the code the same. As we just saw, testing can help us make sure that the functionality does not change as we are refactoring the code but how would we know that the code is getting faster and faster? Measuring software performance is called *profiling*. For profiling, we will rely on functionality that is built into the Jupyter notebook, a notebook magic command called `%timeit` that measures the runtime of a line of code. Depending on how long it takes, Jupyter may decide to run it multiple times so that it can gather some more data and calculate statistics of the performance of the software (i.e., what are the mean and variance of the runtime when the code is run multiple times). Let's look at a simple example of profiling and refactoring. Consider a function that takes in a sequence of numbers and calculates their average:

```
def average(numbers):
    total = 0
    for number in numbers:
        total += number
    return total / len(numbers)
```

We can test this code to make sure that it does what we would expect in a few simple cases:

```
def test_average():
    assert average([1,1,1]) == 1
    assert average([1,2,3]) == 2
    assert average([2,2,3,3]) == 2.5
```

```
test_average()
```

Nothing happens—it seems that the code does what it is expected to do in these cases. As an exercise, you could add tests to cover more use cases and improve the code. For example, what happens if an empty list is passed as input? What should happen? Next, let's time its execution. To call the `%timeit` magic, we put this command at the beginning of a line of code that includes a call to the function that we are profiling:

```
%timeit average([1,2,3,4,5,6,7,8,9,10])
```

```
463 ns ± 15.5 ns per loop (mean ± std. dev. of 7 runs, 1,000,000 loops each)
```

This is pretty fast, but could we find a way to make it go even faster? What if I told you that Python has a built-in function called sum that takes a sequence of numbers and returns its sum? Would this provide a faster implementation than our loop-based approach? Let's give it a try:

```
def average(numbers):
    return sum(numbers) / len(numbers)
```

First, we confirm that the test still works as expected:

```
test_average()
```

Looks like it does! Let's see if it is also faster:

```
%timeit average([1,2,3,4,5,6,7,8,9,10])
```

```
254 ns ± 10.2 ns per loop (mean ± std. dev. of 7 runs, 1,000,000 loops each)
```

This is great! Using the built-in function provides a more than two-fold speedup in this case. Another factor that will often interest us in profiling code is how performance changes as the size of the input changes, or its *scaling* (this is a measure of the efficiency of the code). We can measure the scaling performance of the function by providing increasingly larger inputs and calling the `%timeit` magic on each of these:

```
for factor in [1,10,100,1000]:
    %timeit average([1,2,3,4,5,6,7,8,9,10] * factor)
```

```
318 ns ± 3.08 ns per loop (mean ± std. dev. of 7 runs, 1,000,000 loops each)
```

```
1.06 µs ± 131 ns per loop (mean ± std. dev. of 7 runs, 1,000,000 loops each)
```

```
5.28 µs ± 78.4 ns per loop (mean ± std. dev. of 7 runs, 100,000 loops each)
```

```
52.2 µs ± 937 ns per loop (mean ± std. dev. of 7 runs, 10,000 loops each)
```

In this case, the runtime grows approximately linearly with the number of items in the input (each subsequent run is approximately ten times slower than the previous run). Measuring scaling performance will be important in cases where you are writing code and testing it on small inputs, but intend to use it on much larger amounts of data; for example, if you are writing a routine that compares pairs of voxels in the brain to each other, things can quickly get out of hand when you go from a small number of voxels to the entire brain!

6.5   Summary

If one of your objectives in reading this book is to become a more productive data scientist, bookmark this chapter and revisit it. It likely contains the set of tools that will bring you closest to this goal. As you work through the ideas in the following chapters and start experimenting with them, we believe that you will see the importance of a set of reliable and comfortable-to-use tools for editing, debugging, testing, and profiling your code. This set of simple ideas forms the base layer of effective data science for any kind of data, particularly with complex and large data sets like the neuroscience-based ones that we will start exploring in subsequent chapters.

6.6   Additional Resources

Other resources for profiling code include a line-by-line performance profiler, originally written by Robert Kern (and available as `line_profiler`[1]).

Another aspect of performance that you might want to measure/profile is memory use. This can be profiled using `memory_profiler`.[2]

1. https://github.com/pyutils/line_profiler
2. https://pypi.org/project/memory-profiler/

7

Sharing Code with Others

Collaboration is an important part of science. It is also fun. In the previous chapters, you incrementally learned how to write code in Python and how to improve it through the processes of debugging, testing, and profiling. In the next few chapters, you will learn more and more about the various ways that you can write software to analyze neuroimaging data. While you work through these ideas, it would be good to keep in mind how you will work with others to use these ideas in the collaborative process that you will undertake. At the very least, for your own sake, and the sake of reproducible research, you should be prepared to revisit the code that you created and to keep working with it seamlessly over time. The principles that will be discussed here apply to collaborations with others, as well as with your closest collaborator (and the one who is hardest to reach): yourself from six months ago. Ultimately, we will also discuss ways in which your code can be used by complete strangers. This would provide the ultimate proof of its reproducibility and also possess greater impact than if you were the only one using it.

7.1   What Should Be Shareable?

While we love Jupyter notebooks as a way to prototype code and to present ideas, the notebook format does not, by itself, readily support the implementation of reusable code—functions that can be brought into many different contexts and executed on many different data sets—or code that is easy to test. Therefore, once you are done prototyping an idea for analysis in the notebook, we usually recommend moving your code into Python files (with the extension .py) from which you can import your code into your work environment (for example, into a notebook). There are many ways to organize your code in a way that will facilitate its use by others. Here, we will advocate for a particular organization that will also facilitate the emergence of reusable libraries of code that you can work on with others, and that follow the conventions of the Python language broadly.

However, not every analysis that you do on your data needs to be easy for others to use. Ideally, if you want your work to be reproducible, it will be possible for others to run your code and get the same results. But this is not what we are talking about here. In the course of your work, you will sometimes find that you are creating pieces of code that are useful for you in more than one place, and may also be useful to others, e.g., collaborators in your lab or other researchers in your field. These pieces of code deserve to be written and shared in a manner that others can easily adopt into their work. To do so, the code needs to be packaged into a library. Here, we will look at the nuts and bolts of doing that, by way of a simplified example.

7.2   From Notebook to Module

In the course of our work on the analysis of some MRI data, we wrote the following code in a script or in a Jupyter notebook:

```python
from math import pi
import pandas as pd

blob_data = pd.read_csv('./input_data/blob.csv')

blob_radius = blob_data['radius']

blob_area = pi * blob_radius ** 2
blob_circ = 2 * pi * blob_radius

output_data = pd.DataFrame({"area":blob_area, "circ":blob_circ})
output_data.to_csv('./output_data/blob_properties.csv')
```

Later on in the book (in chapter 9) we will see exactly what Pandas does. For now, suffice it to say that it is a Python library that knows how to read data from comma-separated value (CSV) files, and how to write this data back out. The math module is a built-in Python module that contains (among many other things) the value of π stored as the variable pi.

Unfortunately, this code is not very reusable, even while the results may be perfectly reproducible (provided the input data is accessible). This is because it mixes file input and output with computations, and different computations with each other (e.g., the computation of area and circumference). Good software engineering strives toward *modularity* and *separation of concerns*. One part of your code should be doing the calculations, and another part of the code should be the one that reads and munges the data, yet other functions should visualize the results or produce statistical summaries.

Our first step is to identify what are reusable components of this script and to move these components into a module. For example, in this case the calculation of area and circumference seem like they could each be (separately) useful in many different contexts. Let's isolate them and rewrite them as functions:

```python
from math import pi
import pandas as pd


def calculate_area(r):
    area = pi * r **2
    return area
```

```python
def calculate_circ(r):
    circ = 2 * pi * r
    return circ

blob_data = pd.read_csv('./input_data/blob.csv')
blob_radius = blob_data['radius']
blob_area = calculate_area(blob_radius)
blob_circ = calculate_circ(blob_radius)

output_data = pd.DataFrame({"area":blob_area, "circ":blob_circ})
output_data.to_csv('./output_data/blob_properties.csv')
```

In the next step, we might move these functions out into a separate file; let's call this file geometry.py and document what each function does:

```python
from math import pi

def calculate_area(r):
    """
    Calculates the area of a circle.

    Parameters
    ----------
    r : numerical
        The radius of a circle

    Returns
    -------
    area : numerical
        The calculated area
    """
    area = pi * r **2
    return area


def calculate_circ(r):
    """
    Calculates the circumference of a circle.

    Parameters
    ----------
    r : numerical
        The radius of a circle

    Returns
    -------
    circ : float or array
        The calculated circumference
    """
    circ = 2 * pi * r
    return circ
```

Documenting these functions will help you, and others understand how the function works. At the very least having a one-sentence description, and detailed descriptions of the functions input parameters and outputs or returns, is helpful. You might recognize that in this case, the docstrings are carefully written to comply with the NumPy docstring guide that we told you about in chapter 5.

7.2.1   Importing and Using Functions

Before we continue to see how we will use the `geometry` module that we created, we need to know a little bit about what happens when you call `import` statements in Python. When you call the `import geometry` statement, Python starts by looking for a file called `geometry.py` in your present working directory.

That means that if you saved `geometry.py` alongside your analysis script, you can now rewrite the analysis script as

```
import geometry as geo
import pandas as pd

blob_data = pd.read_csv('./input_data/blob.csv')
blob_radius = blob_data['radius']
blob_area = geo.calculate_area(blob_radius)
blob_circ = geo.calculate_circ(blob_radius)
output_data = pd.DataFrame({"area":blob_area, "circ":blob_circ})
output_data.to_csv('./output_data/blob_properties.csv')
```

This is already good, because now you can import and reuse these functions across many different analysis scripts without having to copy this code everywhere. You have transitioned this part of your code from a one-off notebook or script to a module. Next, let's see how to transition from a module to a library.

7.3   From Module to Package

Creating modular code that can be reused is an important first step, but so far we are are limited to using the code in the `geometry` module only in scripts that are saved alongside this module. The next level of reusability is to create a library, or a *package*, that can be installed and imported across multiple different projects.

Again, let's consider what happens when `import geometry` is called. If there is no file called `geometry.py` in the present working directory, the next thing that Python will look for is a Python package called `geometry`. What is a Python package? It is a folder that has a file called `__init__.py`. This can be imported just like a module, so if the folder is in your present working directory, importing it will execute the code in `__init__.py`. For example, if you were to put the functions you previously had in `geometry.py` in `geometry/__init__.py` you could import them from there, so long as you are working in the directory that contains the `geometry` directory.

More typically, a package might contain different modules that each have some code. For example, we might organize the package like this:

```
.
└── geometry
    ├── __init__.py
    └── circle.py
```

The code that we previously had in geometry.py is now in the circle.py module of the geometry package. To make the names in circle.py available to us we can import them explicitly like this:

```
from geometry import circle
circle.calculate_area(blob_radius)
```

Or we can have the __init__.py file import them for us, by adding this code to the __init__.py file:

```
from .circle import calculate_area, calculate_circ
```

This way, we can import our functions like this:

```
from geometry import calculate_area
```

This also means that if we decide to add more modules to the package, the __init__.py file can manage all the imports from these modules. It can also perform other operations that you might want to do whenever you import the package. Now that you have your code in a package, you will want to install the code in your machine, so that you can import the code from anywhere on your machine (not only from this particular directory) and eventually also so that others can easily install it and run it on their machines.

To do so, we need to understand one more thing about the import statement. If import cannot find a module or package locally in the present working directory, it will proceed to look for this name somewhere in the Python *path*. The Python path is a list of file system locations that Python uses to search for packages and modules to import. You can see it (and manipulate it!) through Python's built-in sys library.

```
import sys
print(sys.path)
```

If we want to be able to import our library regardless of where in our file system we happen to be working, we need to copy the code into one of the file system locations that are stored in this variable. But not so fast! To avoid making a mess of our file system, let's instead let Python do this for us. The `setuptools`[1] library, part of the Python standard library that ships with the Python interpreter, is intended specifically for packaging and setup of this kind. The main instrument for `setuptools` operations is a file called `setup.py` file, which we will look at next.

7.4   The Setup File

By the time you reach the point where you want to use the code that you have written across multiple projects, or share it with others for them to use in their projects, you will also want to organize the files in a separate directory devoted to your library:

```
.
└── geometry
    ├── geometry
    │   ├── __init__.py
    │   └── circle.py
    └── setup.py
```

Notice that we have two directories called `geometry`: the top-level directory contains both our Python package (in this case, the `geometry` package) as well as other files that we will use to organize our project. For example, the file called `setup.py` is saved in the top-level directory of our library. This is a file that we use to tell Python how to setup our software and how to install it. Within this file, we rely on the Python standard library setuptools module to do a lot of the work. The main thing that we will need to do is to provide setuptools with some metadata about our software and some information about the available packages within our software.

For example, here is a minimal setup file:

```python
from setuptools import setup, find_packages

with open("README.md", "r") as fh:
    long_description = fh.read()

setup(
    name="geometry",
    version="0.0.1",
    author="Ariel Rokem",
    author_email="author@example.com",
    description="Calculating geometric things",
    long_description=long_description,
    long_description_content_type="text/markdown",
    url="https://github.com/arokem/geometry",
    packages=find_packages(),
    classifiers=[
        "Programming Language :: Python :: 3",
        "License :: OSI Approved :: MIT License",
```

---

1. https://setuptools.readthedocs.io/en/latest/

```
        "Operating System :: OS Independent",
        "Intended Audience :: Science/Research",
        "Topic :: Scientific/Engineering"
    ],
    python_requires='>=3.8',
    install_requires=["pandas"]
)
```

The core of this file is a call to a function called `setup`. This function has many different options.[2] One of these options is `install`, which would take all the steps needed to properly install the software in the right way into your Python path. This means that once you are done writing this file and organizing the files and folders in your Python library in the right way, you can call

```
$ python setup.py install
```

to install the library into your Python path, in such as way that calling `import geometry` from anywhere in your file system will be able to find this library and use the functions stored within it. Next, let's look at the contents of the file section by section.

7.4.1 Contents of a `setup.py` file

The first thing that happens in the `setup.py` (after the `import` statements at the top) is that a long description is read from a README file. If you are using GitHub to track the changes in your code and to collaborate with others (as described in Section 3), it is a good idea to use the markdown format for this. This is a text-based format that uses the `.md` extension. GitHub knows how to render these files as nice-looking web pages, so your README file will serve multiple different purposes. Let's write something informative in this file. For the `geometry` project it can be something like this:

```
# geometry

This is a library of functions for geometric calculations.

# Contributing

We welcome contributions from the community. Please create a fork of
the project on GitHub and use a pull request to propose your
changes. We strongly encourage creating an issue before starting
to work on major changes, to discuss these changes first.

# Getting help

Please post issues on the project GitHub page.
```

---

2. https://setuptools.readthedocs.io/en/latest/setuptools.html#command-reference

The second thing that happens is a call to the setup function. The function takes several keyword arguments. The first few ones are general metadata about the software:

```
name="geometry",
author="Ariel Rokem",
author_email="author@example.com",
description="Calculating geometric things",
long_description=long_description,
```

The next one makes sure that the long description gets properly rendered in web pages describing the software (e.g., in the Python package index, PyPi[3]; more about that subsequently).

```
long_description_content_type="text/markdown",
```

Another kind of metadata are classifiers that are used to catalog the software within PyPI so that interested users can more easily find them.

```
classifiers=[
    "Programming Language :: Python :: 3",
    "License :: OSI Approved :: MIT License",
    "Operating System :: OS Independent",
    "Intended Audience :: Science/Research",
    "Topic :: Scientific/Engineering"
],
```

In particular, note the license classifier. If you intend to share the software with others, please provide a license that defines how others can use your software. You do not have to make anything up. Unless you have a good understanding of the legal implications of the license, it is best to use a standard OSI-approved license.[4] If you are interested in publicly providing the software in a manner that would allow anyone to do whatever they want with the software, including in commercial applications, the MIT license is not a bad way to go.

The next item is the version of the software. It is a good idea to use the semantic versioning conventions,[5] which communicate to potential users how stable the software is, and whether changes have happened that dramatically change how the software operates:

```
version="0.0.1",
```

The next item points to a URL for the software. For example, the GitHub repository for the software.

3. https://pypi.org/
4. https://opensource.org/licenses
5. https://semver.org/

```
url="https://github.com/arokem/geometry",
```

The next item calls a `setuptools` function that automatically traverses the file system in this directory and finds the packages and subpackages that we have created.

```
packages=find_packages(),
```

If you would rather avoid that, you can also explicitly write out the names of the packages that you would like to install as part of the software. For example, in this project, it could be:

```
packages=['geometry']
```

The last two items define the dependencies of the software. The first is the version of Python that is required for the software to run properly. The other is a list of other libraries that are imported within our software and that need to be installed before we can install and use our software (i.e., they are not part of the Python standard library). In this case, only the Pandas library. Though we have not started using it in our library code yet, we might foresee a need for it at a later stage, for example, when we add code that reads and writes data files. For now, we have added it here as an example.

```
python_requires='>=3.8',
install_requires=["pandas"]
```

7.5   A Complete Project

At this point, our project is starting to take shape. The file system of our library should look something like this:

```
.
└── geometry
    ├── LICENSE
    ├── README.md
    ├── geometry
    │   ├── __init__.py
    │   └── circle.py
    └── setup.py
```

In addition to the `setup.py` file, we have added the `README.md` file, as well as a `LICENSE` file that contains the license we have chosen to apply to our code. At this point, the project has everything it needs to have for us to share it widely and for others to start using it. We can add all these files and then push this into a repository on GitHub. Congratulations! You have created a software project that can easily be installed on your machine, and on other computers. What next? There are a few further steps you can take.

7.5.1   Testing and Continuous Integration

We already looked at software testing in chapter 6. As we mentioned there, tests are particularly useful if they are automated and run repeatedly. In the context of a well-organized Python project, that can be achieved by including a test module for every module in the library. For example, we might add a `tests` package within our `geometry` package

```
.
└── geometry
    ├── LICENSE
    ├── README.md
    ├── geometry
    │   ├── __init__.py
    │   ├── tests
    │   │   ├── __init__.py
    │   │   └── test_circle.py
    │   └── circle.py
    └── setup.py
```

where `__init__.py` is an empty file, signaling that the `tests` folder is a package as well, and the `test_circle.py` file may contain a simple set of functions for testing different aspects of the code. For example the code

```python
from geometry.circle import calculate_area
from math import pi

def test_calculate_area():
    assert calculate_area(1) == pi
```

will test that the `calculate_area` function does the right thing for some well-understood input.

To reduce the friction that might prevent us from running the tests often, we can take advantage of systems that automate the running of tests as much as possible. The first step is to use software that runs the tests for us. These are sometimes called *test harnesses*. One popular test harness for Python is Pytest.[6] When it is called within the source code of your project, the Pytest test harness identifies functions that are software tests by looking for them in files whose names start with `test_` or end with `_test.py`, and by the fact that the functions themselves are named with names that start with `test_`. It runs these

---

6. https://docs.pytest.org/

functions and keeps track of the functions that pass the test; i.e., do not raise errors, and those that fail; i.e., do raise errors.

Another approach that can automate your testing, even more, is called *continuous integration*. In this approach, the system that keeps track of versions of your code (e.g., the GitHub website) also automatically runs all of the tests that you wrote every time that you make changes to the code. This is a powerful approach in tandem with the collaborative tools that we described in chapter 3 because it allows you to identify the exact set of changes that changed a passing test into a failing test and to alert you to this fact. In the collaborative pattern that uses pull requests, the tests can be run on the code before it is ever integrated into the main branch, allowing contributors to fix changes that cause test failures before they are merged. Continuous integration is implemented in GitHub through a system called GitHub Actions. We will not go further into the details of this system here, but you can learn about it through the online documentation.[7]

7.5.2 Documentation

If you follow the instructions we provided previously, you will have already made the first step toward documenting your code by including docstrings in your function definitions. A further step is to write more detailed documentation and make the documentation available together with your software. A system that is routinely used across the Python universe is Sphinx.[8] It is a rather complex system for generating documentation in many different formats, including a PDF manual, but also a neat-looking website that includes your docstrings and other pages you write. Creating such a website can be a worthwhile effort, especially if you are interested in making your software easier for others to find and to use.

7.6 Summary

If you go through the work described in this section, making the software that you write for your use easy to install and openly available, you will make your work easier to reproduce, and also easier to extend. Other people might start using it. Inevitably, some of them might run into bugs and issues with the software. Some of them might even contact you to ask for help with the software, either through the "Issues" section of your GitHub repository or via email. This could lead to fruitful collaborations with other researchers who use your software. On the one hand, it might be a good idea to support the use of your software. One of our goals as scientists is to have an impact on the understanding of the universe and the improvement of the human condition, and supported software is more likely to have such an impact. Furthermore, with time, users can become developers of the software, initially by helping you expose errors that may exist in the code, and ultimately by contributing new

---

7. https://github.com/features/actions
8. https://www.sphinx-doc.org/en/master/

features. Some people have made careers out of building and supporting a community of users and developers around software they write and maintain. On the other hand, that might not be your interest or your purpose, and it does take time away from other things. Either way, you can use the README portion of your code to communicate the level of support that others might expect. It is perfectly fine to let people know that the software is provided openly, but that it is provided with no assurance of any support.

7.6.1   Software Citation and Attribution

Researchers in neuroimaging are used to the idea that when we rely on a paper for the findings and ideas in it, we cite it in our research. We are perhaps less accustomed to the notion that software that we use in our research should also be cited. In recent years, there has been an increased effort to provide ways for researchers to cite software, and also for researchers who share their software to be cited. One way to make sure that others can cite your software is to make sure that your software has a digital object identifier (or DOI). A DOI is a string of letters and numbers that uniquely identifies a specific digital object, such as an article or data set. It is used to keep track of this object and identify it, even if the web address to the object may change. Many journals require that a DOI be assigned to a digital object before that object can be cited so that the item that is cited can be found even after some time. This means that to make your software citeable, you will have to mint a DOI for it. One way to do that is through a service administered by the European Council for Nuclear Research (CERN) called Zenodo.[9] Zenodo allows you to upload digital objects—the code of a software library, for example—into the website, and then provides a DOI for them. It even integrates with the GitHub website to automatically provide a separate DOI for every version of a software project. This is also a reminder that when you are citing a software library that you use in your research, make sure to refer to the specific version of the software that you are using, to make sure that others reading your article can reproduce your work.

7.7   Additional Resources

Packaging and distributing Python code involves putting together some rather complex technical pieces. In response to some of the challenges, the Python community put together the Python Packaging Authority (PyPA) website,[10] which explains how to package and distribute Python code. Their website is a good resource to help you understand some of the machinery that we explained previously, and also to update it with the most recent best practices.

---

9. https://zenodo.org
10. https://www.pypa.io/en/latest/

In Section 4.1 you learned about the Conda package manager. A great way to distribute scientific software using conda is provided through the Conda Forge[11] project, which supports members of the community by providing some guidance and recipes to distribute software using Conda.

Jake VanderPlas wrote a very useful blog post[12] on the topic of scientific software licensing.

Developing an open-source software project can also become a complex social, technical, and even legal challenge. A (free!) book you might want to look at if you are contemplating taking it on is Producing Open Source Software[13] by Karl Fogel, which will take you through everything from naming an open-source software project to legal and management issues such as licensing, distribution, and intellectual property rights.

11. https://conda-forge.org/
12. https://www.astrobetter.com/blog/2014/03/10/the-whys-and-hows-of-licensing-scientific-code/
13. https://producingoss.com/

PART III
Scientific Computing

8

The Scientific Python Ecosystem

Python has an excellent ecosystem of scientific computing tools. What do we mean when we say *scientific computing*? There is no one definition, but we roughly mean those parts of computing that allow us to work with scientific data. Data that represent images or time series from neuroimaging experiments definitely fall under that definition. It is also often called *numerical computing* because these data are often stored as numbers. A common structure for numerical computing is that of an array. An array is an ordered collection of items that are stored such that a user can access one or more items in the array. The NumPy[1] Python library provides an implementation of an array data structure that is both powerful and lightweight.

What do we mean by *ecosystem*? This means that many different individuals and organizations develop software that interoperates to perform the various scientific computing tasks that they need to perform. Tools compete for users and developers' attention and mindshare, but more often they might coordinate. No one entity manages this coordination. In that sense, the collection of software tools that exists in Python is more like an organically evolving ecosystem than a centrally managed organization. One place in which the ecosystem comes together as a community is an annual conference called "Scientific Computing in Python,"[2] which takes place in Austin, Texas every summer.

8.1 Numerical Computing in Python

8.1.1 From Naive Loops to n-Dimensional Arrays

In many disciplines, data analysis consists largely of the analysis of tabular data. This means two-dimensional tables where data are structured into rows and columns, with each observation typically taking up a row, and each column representing a single variable. Neuroimaging data analysis also often involves tabular data, and we will spend a lot of time in chapter 9 discussing the tools Python provides for working efficiently with tabular data.

---

1. https://numpy.org/
2. https://conference.scipy.org/

However, one of the notable features of neuroscience experiments is that data sets are often large, and naturally tend to have many dimensions. This means that data from neuroscience experiments are often stored in big tables of data called *arrays*. Much like Python lists or tuples, the items in an array follow a particular order. But in contrast to a list or a tuple, they can have multiple dimensions: arrays can have one dimension (e.g., a time series from a single channel of an electroencephalography (EEG) recording), two dimensions (e.g., all the time series from multiple EEG channels recorded simultaneously, or a single slice from a magnetic resonance imaging (MRI) session), three dimensions (e.g., an anatomical scan of a human brain), four dimensions (e.g., a functional MRI [fMRI] data set with the three spatial dimensions and time points arranged as a series of volumes along the last dimension) or even more.

In all of these cases, the arrays are *contiguous*. That means that there are no holes in an array, where nothing is stored. They are also *homogeneous*, which means that all of the items in an array are of the same numerical data type. That is, if one number stored in the array is a floating point number, all numbers in the array will be represented in their floating point representation. It also means that the dimensions of the array are predictable. If the first channel of the EEG time series has 20,000 time points in it, the second channel would also have 20,000 time points in it. If the first slice of an MRI image has 64×64 voxels, other slices in the MRI image would also have 64×64 voxels.

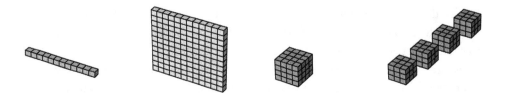

To understand why this is challenging, let's consider a typical fMRI study, in which participants lie in the scanner, while their brain responses are measured. Suppose we have twenty participants, each scanned for roughly 30 minutes, with a repetition time (TR), i.e., the duration of acquisition of each fMRI volume, of 1 second. If the data are acquired at an isotropic spatial resolution of 2 mm (i.e., each brain *voxel*, or three-dimensional pixel is 2 mm along each dimension), then the resulting data set might have approximately $20 \times 1,800 \times 100 \times 100 \times 100 = 36$ billion observations. That is a lot of data! Moreover, each subject's data has a clear four-dimensional structure; i.e., the three spatial dimensions, plus time. If we wanted to, we could also potentially represent subjects as the fifth dimension, though that involves some complications because at least initially, different subjects' brains will not be aligned with one another; we would need to spatially register them for that (we will learn more about that in chapter 16).

This means that a single subject's data would look a little bit like the four-dimensional example that we just discussed.

We will typically want to access this data in pretty specific ways. That is, rather than applying an operation to every single voxel in the brain at every single point in time, we usually want to pull out specific slices of the data and only apply an operation to those slices. For example, say we are interested in a voxel in the amygdala. How would we access only that voxel at every time point?

A very naive approach that we could implement in pure Python would be to store all our data as a nested series of Python lists: each element in the first list would be a list containing data for one time point; each element within the list for each time point would itself be a list containing the two-dimensional slices at each x-coordinate; and so on. Then we could write a series of four nested for-loops to sequentially access every data point (or voxel) in our array. For each voxel we inspect, we could then determine whether the voxel is one we want to work with, and if so, apply some operation to it.

Basically, we would have something like this (note that this is just pseudocode, not valid Python code, and you would not be able to execute this snippet):

```
for t in time:
    for x in t:
        for y in x:
            for z in y:
                if z is in amygdala:
                    apply_my_function(z)
```

Something similar to this example would probably work just fine. But it would be very inefficient, and might take a while to execute. Even if we only want to access a single amygdala voxel at each time point (i.e., 1,800 data points in total, if we stick with our previous resting state example), our naive looping approach still requires us to inspect every single one of the 1.8 billion voxels in our data set—even if all we are doing for the vast majority of voxels is deciding that we do not actually need to use that voxel in our analysis! This is, to put it mildly, a very bad thing.

Computer scientists have developed a notation for more precisely describing how much of a bad thing it is: Big O notation. This gives us a formal way of representing the time complexity of an algorithm. In Big O notation, the code in the previous example runs in $O(n^4)$ time, meaning for any given dimension size n (we assume for simplicity that n is the same for all dimensions), we need to perform on the order of n^4 computations. To put it in perspective, just going from $n = 2$ to $n = 4$ means that we go from $2^4 = 16$ to $4^4 = 256$ operations. You can see how this might start to become problematic as n gets large.

The lesson we should take away from this is that writing naive Python loops just to access the values at a specific index in our data set is a bad idea.

8.1.2 Toward a Specialized Array Structure

What about list indexing? Fortunately, since we have already covered list indexing in Python, we have a better way to access the amygdala data we want. We know that the outer dimension in our nested list represents time, and we also know that the spatial index of the amygdala voxel we want will be constant across time points. This means that instead of writing a four-level nested for-loop, we can do the job in just a single loop. If we know that the index of the amygdala voxel we want in our three-dimensional image is [20, 18, 32], then the following code suffices (again, note that this code cannot be executed):

```python
amygdala_values = []

for t in dataset:
    amygdala_values.append(t[20][18][32])
```

That is a big improvement, in terms of time complexity. We have gone from an $O(n^4)$ to an $O(n)$ algorithm. The time our new algorithm takes to run is now linear in the number of time points in our data set, rather than growing as a polynomial function of it.

Unfortunately, our new approach still is not great. One problem is that performance still will not be so good, because Python is not naturally optimized for the kind of operation we just saw; i.e., indexing repeatedly into nested lists. For one thing, Python is a dynamically typed language, so there will be some overhead involved in accessing each individual element (because the interpreter has to examine the object and determine its type), and this can quickly add up if we have millions or billions of elements.

The other big limitation is that, even if lists themselves were super fast to work with, the Python standard library provides limited functionality for doing numerical analysis on them. There are a lot of basic numerical operations we might want to apply to our data that would be really annoying to write using only the built-in functionality of Python. For example, Python lists do not natively support matrix operations, so if we wanted to multiply one matrix (represented as a list of lists) by another, we would probably need to write out a series of summations and multiplications ourselves inside for-loops. And that would be completely impractical given how basic and common an operation matrix multiplication is (not too important for our point now, but if you are wondering what a matrix multiplication is, you can see a formal definition in chapter 13).

The bottom line is that, as soon as we start working with large, or highly structured data sets—as is true of most data sets in neuroimaging—we are going to have to look beyond what Python provides us in its standard library. Lists just are not going to cut it; we need some other kind of data structure, one that is optimized for numerical analysis on large, n-dimensional data sets. This is where NumPy comes into play.

8.2   Introducing NumPy

Undoubtedly, NumPy is the backbone of the numerical and scientific computing stack in Python; many of the libraries we will cover in this book (Pandas, Scikit-learn, etc.) depend on it internally. It provides many data structures optimized for efficient representation and manipulation of different kinds of high-dimensional data, as well as an enormous range of numerical tools that help us work with those structures.

8.2.1 *Importing* Numpy

Recall that the default Python namespace contains only a small number of built-in functions. To use any other functionality, we need to explicitly import it into our namespace. Let's do that for NumPy. The library can be imported using the `import` keyword. We use `import numpy as np` to create a short name for the library that we can refer to. Everything that is in `numpy` will now be accessible through the short name `np`. Importing the library specifically as `np` is not a requirement, but it is a very strongly held convention, and when you read code that others have written you will find that this the form that is very often used.

```
import numpy as np
```

8.2.2 *Data is Represented in Arrays*

The core data structure in NumPy is the n-dimensional array (or `ndarray`). As the name suggests, an `ndarray` is an array with an arbitrary number of dimensions. Unlike Python lists, NumPy arrays are homogeneously typed, meaning every element in the array has to have the same data type. You can have an array of floats, or an array of integers, but you cannot have an array that mixes floats and integers. This is exactly what we described previously: a homogeneous dense table that holds items in it.

8.2.3 *Creating an* ndarray

Like any other Python object, we need to initialize an `ndarray` before we can do anything with it. Luckily, NumPy provides us with several ways to create new arrays. Let's explore a couple.

INITIALIZING AN ARRAY FROM AN EXISTING LIST

Let's start by constructing an array from existing data. Assume we have some values already stored in a native Python iterable object; typically, this is a list (*iterable* means that it is an object that we can iterate over with a for-loop, see Section 5.5.2). If we want to convert that object to an `ndarray` so that we can perform more efficient numerical operations on it, we can pass the object directly to the `np.array()` function. In a relatively simple case, the input to the `np.array` function is a sequence of items, a list that holds some integers.

```
list_of_numbers = [1, 1, 2, 3, 5]
arr1 = np.array(list_of_numbers)
print(arr1)
```

```
[1 1 2 3 5]
```

The result still looks a lot like a list of numbers, but let's look at some of the attributes that this variable has that are not available in lists.

```
print("The shape of the array is", arr1.shape)
print("The dtype of the array is", arr1.dtype)
print("The size of each item is", arr1.itemsize)
print("The strides of the array is", arr1.strides)
```

```
The shape of the array is (5,)
The dtype of the array is int64
The size of each item is 8
The strides of the array is (8,)
```

The shape of an array is its dimensionality. In this case, the array is one-dimensional (like the single channel of EEG data in the previous example). All of the items in this array are integers, so we can use an integer representation as the dtype of the array. This means that each element of the array takes up 8 bytes of memory. Using 64-bit integers, we can represent the numbers from -9223372036854775808 to 9223372036854775807, and each one would take up 8 bytes of memory in our computer (or 64 bits). This also explains the strides. Strides are the number of bytes in memory that we would have to skip to get to the next value in the array. Because the array is packed densely in a contiguous segment of memory, to get from the value 1 to the value 2, for example, we have to skip 8 bytes.

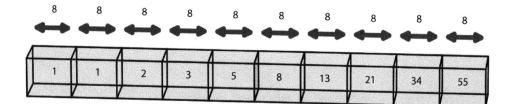

Remember that the data we would like to represent in arrays can have multiple dimensions. Fortunately, NumPy is pretty good about figuring out what kind of array we want based on the input data; for example, if we pass in a list of lists, where each of the inner lists has the same number of elements, NumPy will infer that we want to construct a two-dimensional array.

```
list_of_lists = [[1, 1, 2, 3, 5], [8, 13, 21, 34, 55]]
arr2 = np.array(list_of_lists)

print(arr2)
print("The shape of this array is", arr2.shape)
print("The dtype of this array is", arr2.dtype)
```

```
print("The size of each item is", arr2.itemsize)
print("The strides of this array are", arr2.strides)
```

```
[[ 1  1  2  3  5]
 [ 8 13 21 34 55]]
The shape of this array is (2, 5)
The dtype of this array is int64
The size of each item is 8
The strides of this array are (40, 8)
```

Now, the shape of the array is two items: the first is the number of lists in the array (i.e., two) and the second is the number of items in each list (five). You can also think of that as rows (two rows) and columns (five columns), because each item in the first row of the array has an item matching it in terms of its column in the second row. For example, the number 2 in the first row is equivalent to the number 21 in the second row in that they are both the third item in their particular rows. Or, in other words, they share the third column in the array. The items are still each 8 bytes in size, but now there are two items in the strides: (40, 8). The first item tells us how many bytes we have to move to get from one row to the next row (e.g., from the 2 in the first row to the 21 in the second row). That is because the part of the computer's memory in which the entire array is stored is contiguous and linear. Here, we have drawn this contiguous bit of memory, such that the two rows of the array are laid out side-by-side.

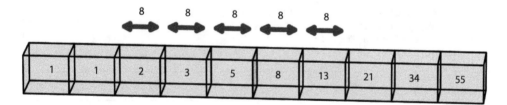

The first time you see a printed representation of a NumPy array, as in the output shown here, it might look a little confusing. But the visual representation closely resembles what we would see if we were to print the nested list in Python. In the previous two-dimensional case, the elements of the outer list are the values along the first dimension, and each inner list gives the values along the second dimension. This ends up giving us a nice tabular representation, where, at least for small arrays, we can just read off the values from the output. For example, the first row of the array contains the values [1, 1, 2, 3, 5]; the second column of the array contains the values [1, 13]; and so on.

Exercise

Extend the above principle into three dimensions: create a list of list of lists, initialize an array from the list, and print into the screen. Make sure you understand how the values displayed visually map onto the dimensions of the array.

Alternatively, we can construct a new array from scratch and fill it with some predetermined value—most commonly zero. We can do this easily in NumPy using the conveniently named `np.zeros` function. In many practical applications, we can think of this as an *empty* array (though technically we could also create a truly empty array that has no assigned values using the `np.empty` function).

The `zeros` function takes a mandatory shape tuple as its first argument; this specifies the dimensions of the desired array.

```python
arr_2d = np.zeros((5, 10))
print("The shape of this array is", arr_2d.shape)
print("The dtype of this array is", arr_2d.dtype)
print("The size of each item is", arr_2d.itemsize)
print("The strides of this array are", arr_2d.strides)

arr_3d = np.zeros((2, 4, 8))
print("The shape of this array is", arr_3d.shape)
print("The dtype of this array is", arr_3d.dtype)
print("The size of each item is", arr_3d.itemsize)
print("The strides of this array are", arr_3d.strides)
```

```
The shape of this array is (5, 10)
The dtype of this array is float64
The size of each item is 8
The strides of this array are (80, 8)
The shape of this array is (2, 4, 8)
The dtype of this array is float64
The size of each item is 8
The strides of this array are (256, 64, 8)
```

Exercise

Can you explain the outputs of the previous code cell? What does the first item in the `strides` represent in each case?

8.2.4 Neuroscience Data Are Stored in Arrays

As we described previously, many different kinds of neuroscience data are naturally stored in arrays. To demonstrate this, let's look at some neuroscience data; we have stored some blood oxygenation level dependent (BOLD) data from a functional MRI (fMRI) experiment in a NumPy array, and you can load it using our the `load_data` utility function from the `nsdlib` software library:

```python
from ndslib import load_data
```

This load_data function takes as arguments the name of the data set that we would like to download and optionally also the name of a file that we would like to save the data into (for reuse). The npy file format is a format that was developed specifically to store NumPy arrays and that is all that it knows how to store. When you load one of these files, you can assign into a variable the contents of the array that was stored in the file.

```
bold = load_data("bold_numpy", fname="bold.npy")

print("The shape of the data is", bold.shape)
print("The dtype of the data is", bold.dtype)
print("The itemsize of the data is", bold.itemsize)
print("The strides of the data is", bold.strides)
```

```
The shape of the data is (64, 64, 25, 180)
The dtype of the data is float64
The itemsize of the data is 8
The strides of the data is (8, 512, 32768, 819200)
```

These BOLD signal data are a little bit more complicated than the arrays you have seen so far. It has four dimensions: the first two dimensions are the in-plane dimensions of each slice: 64×64 voxels. The next dimension is the slice dimension: twenty-five slices. Finally, the last dimension is the time points that were measured in the experiment. The dtype of this array is float64. These floating point numbers are what computers use to represent real numbers. Each one of these items also uses up 64 bits, or 8 bytes, so that is the item size here as well. Just like the shape, there are four strides: the first item in the strides tells how far we need to go to get from one voxel to another voxel in the same row, in the same slice, in the same time point. The second stride (i.e., 512) tells us how far we would have to go to find the same voxel in the same slice and time point, but one row over. That is sixty-four times eight. The next one is about finding the same voxel in a neighboring slice. Finally the large one at the end is about finding the same slice in the next time step. That large number is equal to $64 \times 64 \times 25$ (the number of voxels in each time point) $\times 8$ (the number of bytes per element in the array).

8.2.5   Indexing Is Used to Access Elements of an Array

We have seen how we can create arrays and describe them; now let's talk about how we can get data in and out of arrays. We already know how to index Python lists, and array indexing in Python will look quite similar. But NumPy indexing adds considerably more flexibility and power, and developing array indexing facility is a critical step toward acquiring general proficiency with the package. Let's start simple: accessing individual items in a one-dimensional array looks a lot like indexing into a list (see Section 5.3.1). For example, for the first item in the one-dimensional array we saw earlier, you would do the following:

```
arr1[0]
```

```
1
```

If you want the second element, you would do something like this (do not forget that Python uses zero-based indexing):

```
arr1[1]
```

```
1
```

NumPy arrays also support two other important syntactic conventions that are also found in Python lists: indexing from the end of the array, and slicing the array to extract multiple contiguous values.

To index from the end, we use the minus sign $(-)$:

```
arr1[-2]
```

```
3
```

This gives us the second-from-last value in the array. To slice an array, we use the colon $(:)$ operator, passing in the positions we want to start and end at:

```
arr1[2:5]
```

```
array([2, 3, 5])
```

As in lists, the start index is inclusive and the end index is exclusive (i.e., in the previous example, the resulting array includes the value at position 2, but excludes the one at position 6). We can also omit the start or end indexes, in which case, NumPy will return all positions up to, or starting from, the provided index. For example, to get the first four elements we would do the following:

```
arr1[:4]
```

```
array([1, 1, 2, 3])
```

8.2.6   Indexing in Multiple Dimensions

Once we start working with arrays with more than one dimension, indexing gets a little more complicated. Both because the syntax is a little different, and because there is a lot more we can do with multidimensional arrays.

For example, if you try one of these operations with a two-dimensional array, something funny will happen:

```
arr2[0]
```

```
array([1, 1, 2, 3, 5])
```

It looks like the first item of a two-dimensional array is the entire first row of that array. To access the first element in the first row, we need to use two indices, separated by a comma:

```
arr2[0, 0]
```

```
1
```

One way to think about this is that the first index points to the row and the second index points to the column, so to get the very first item in the array, we need to explicitly ask for the first column in the first row. Consider what happens when you ask for third item in the second row:

```
arr2[1, 2]
```

```
21
```

Under the hood, Python asks how many bytes into the array it would have to go to get that item. It looks at the strides and computes: 1 times the first stride $= 40$, plus 2 times the second stride $= 40 + 16 = 56$. Indeed, the third element of the second row of the array is 7 times 8 bytes into the array:

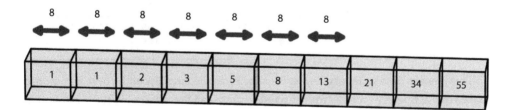

How would you access data in the `bold` array? As an example, let's access the voxel in the very first time point, in the very first slice, and in this slice in the first row and column:

```
bold[0, 0, 0, 0]
```

```
0.0
```

This element of the array is the very corner of the volume, so it is way outside of the part of the data that contains the brain. That is why it contains a value of zero. Instead, we can look at the voxel at the very center of the center slice:

```
bold[32, 32, 12, 0]
```

```
1080.0
```

This voxel has a value that corresponds to the intensity of the magentic resonance image in this voxel at the first time point. If we drop the last index, that would be like asking for the data in all of the time points in that particular voxel:

```
bold[32, 32, 12]
```

```
array([1080., 1052., 1056., 1087., 1146., 1147., 1105., 1064., 1128.,
       1089., 1095., 1049., 1109., 1051., 1074., 1073., 1112., 1086.,
       1090., 1062., 1086., 1023., 1047., 1139., 1065., 1117., 1070.,
       1070., 1089., 1074., 1051., 1034., 1096., 1060., 1096., 1076.,
       1032., 1067., 1030., 1072., 1056., 1069., 1061., 1054., 1072.,
       1072., 1035., 1018., 1116., 1056., 1051., 1084., 1075., 1080.,
       1036., 1022., 1076., 1060., 1031., 1079., 1048., 1002., 1055.,
       1027., 1014., 1006., 1072., 1003., 1026., 1039., 1096., 1078.,
       1025., 1029., 1009., 1065., 1023., 1098., 1045., 1094., 1016.,
       1015., 1027., 1020., 1030., 1049., 1034., 1053., 1018., 1038.,
       1072., 1020., 1007., 1037., 1082., 1050., 1011., 1027.,  972.,
        992.,  984., 1021., 1018., 1029., 1063., 1070., 1033., 1048.,
       1052., 1040., 1007.,  996., 1026., 1011., 1023.,  958.,  982.,
       1008., 1059., 1006., 1012., 1042., 1000., 1066., 1026., 1032.,
       1009., 1022.,  989., 1020., 1078., 1034., 1036.,  976., 1026.,
        960., 1021., 1009., 1049., 1029., 1065., 1002., 1007., 1031.,
       1015., 1044., 1012.,  999., 1002.,  979.,  965., 1011., 1027.,
       1013., 1012., 1022.,  981., 1030., 1061., 1056., 1014.,  957.,
        968., 1015., 1100.,  994., 1000.,  973.,  997.,  980., 1026.,
       1004., 1008.,  988.,  993., 1042., 1016., 1044., 1002., 1006.])
```

This is now a time series of data, corresponding to the intensity of the BOLD response in the voxel at the very center of the central slice, across all time points in the measurement.

8.2.7   Slicing a Dimension with " : "

As you saw before, we can use slicing to choose a subarray from within our array. For example, you can slice out the time series in the central voxel, as we just did:

```
bold[32, 32, 12, :]
```

```
array([1080., 1052., 1056., 1087., 1146., 1147., 1105., 1064., 1128.,
       1089., 1095., 1049., 1109., 1051., 1074., 1073., 1112., 1086.,
       1090., 1062., 1086., 1023., 1047., 1139., 1065., 1117., 1070.,
       1070., 1089., 1074., 1051., 1034., 1096., 1060., 1096., 1076.,
       1032., 1067., 1030., 1072., 1056., 1069., 1061., 1054., 1072.,
       1072., 1035., 1018., 1116., 1056., 1051., 1084., 1075., 1080.,
       1036., 1022., 1076., 1060., 1031., 1079., 1048., 1002., 1055.,
       1027., 1014., 1006., 1072., 1003., 1026., 1039., 1096., 1078.,
       1025., 1029., 1009., 1065., 1023., 1098., 1045., 1094., 1016.,
       1015., 1027., 1020., 1030., 1049., 1034., 1053., 1018., 1038.,
       1072., 1020., 1007., 1037., 1082., 1050., 1011., 1027.,  972.,
        992.,  984., 1021., 1018., 1029., 1063., 1070., 1033., 1048.,
       1052., 1040., 1007.,  996., 1026., 1011., 1023.,  958.,  982.,
       1008., 1059., 1006., 1012., 1042., 1000., 1066., 1026., 1032.,
       1009., 1022.,  989., 1020., 1078., 1034., 1036.,  976., 1026.,
        960., 1021., 1009., 1049., 1029., 1065., 1002., 1007., 1031.,
       1015., 1044., 1012.,  999., 1002.,  979.,  965., 1011., 1027.,
       1013., 1012., 1022.,  981., 1030., 1061., 1056., 1014.,  957.,
        968., 1015., 1100.,  994., 1000.,  973.,  997.,  980., 1026.,
       1004., 1008.,  988.,  993., 1042., 1016., 1044., 1002., 1006.])
```

This is equivalent to dropping the last index. If we want to slice through all of the slices, taking the values of the central voxel in each slice, in the very first time point, we would instead slice on the third (slice) dimension of the array:

```
bold[32, 32, :, 0]
```

```
array([  31.,   52.,  117.,  423.,   77.,  312.,  454.,  806.,  576.,
        581.,  403., 1031., 1080.,  482.,  439.,  494.,  546.,  502.,
        453.,  362.,  385.,  367.,  344.,  169.,   25.])
```

These twenty-five values correspond to the values in the center of each slice in the first time point. We can also slice on multiple dimensions of the array. For example, if we would like to get the values of all the voxels in the central slice in the very first time point, we can do the following:

```
central_slice_t0 = bold[:, :, 12, 0]
```

This is a two-dimensional array, with the dimensions of the slice

```
central_slice_t0.shape
```

```
(64, 64)
```

which we can verify by printing out the values inside the array:

```
print(central_slice_t0)
```

```
[[ 0.   0.   0.  ...  0.   0.   0.]
 [13.   4.   3.  ... 16.   3.   6.]
 [ 6.  13.   6.  ... 13.  26.   7.]
 ...
 [13.  11.  17.  ... 16.   6.   5.]
 [10.   3.   7.  ...  8.   6.   8.]
 [ 7.   7.   7.  ... 20.  16.   4.]]
```

8.2.8   Indexing with Conditionals

One more way of indexing is to to use logical operations, or conditionals. This allows us to choose values from an array, only if they fulfill certain conditions. For example, if we would like to exclude measurements with a value of zero or less, we might perform a logical operation like the following:

```
larger_than_0 = bold>0
```

This creates a new array that has the same shape as the original array, but contains Boolean (True/False) values. Wherever we had a number larger than zero in the original array, this new array will have the value True and wherever the numbers in the original array were zero or smaller, this new array will have the value False. We can use this array to index into the original array. What this does is that it pulls out of the original array only the values for which larger_than_0 is True and reorganizes these values as a long one-dimensional array:

```
bold[larger_than_0]
```

```
array([ 6., 11.,  5., ..., 12.,  4., 14.])
```

This can be useful, for example, if you would like to calculate statistics of the original array, but only above some threshold value. For example, if you would like to calculate the average of the array (using the mean method of the array) with and without values that are zero or less:

```
print(bold.mean())
print(bold[larger_than_0].mean())
```

```
145.360578125
```

```
148.077239538181
```

Exercise

Arrays have a variety of methods attached to them to do calculations. Use the array method `std` to calculate the standard deviation of the `bold` array, and the standard deviation of just values of the array that are between 1 and 100.

8.2.9 Arithmetic with Arrays

In addition to storing data from a measurement, and providing access to the this data through indexing, arrays also support a variety of mathematical operations with arrays. For example, you can perform arithmetic operations between an array and a number. Consider for example what happens when we add a single number to an entire array:

```
bold[32, 32, 12] + 1000
```

```
array([2080., 2052., 2056., 2087., 2146., 2147., 2105., 2064., 2128.,
       2089., 2095., 2049., 2109., 2051., 2074., 2073., 2112., 2086.,
       2090., 2062., 2086., 2023., 2047., 2139., 2065., 2117., 2070.,
       2070., 2089., 2074., 2051., 2034., 2096., 2060., 2096., 2076.,
       2032., 2067., 2030., 2072., 2056., 2069., 2061., 2054., 2072.,
       2072., 2035., 2018., 2116., 2056., 2051., 2084., 2075., 2080.,
       2036., 2022., 2076., 2060., 2031., 2079., 2048., 2002., 2055.,
       2027., 2014., 2006., 2072., 2003., 2026., 2039., 2096., 2078.,
       2025., 2029., 2009., 2065., 2023., 2098., 2045., 2094., 2016.,
       2015., 2027., 2020., 2030., 2049., 2034., 2053., 2018., 2038.,
       2072., 2020., 2007., 2037., 2082., 2050., 2011., 2027., 1972.,
       1992., 1984., 2021., 2018., 2029., 2063., 2070., 2033., 2048.,
       2052., 2040., 2007., 1996., 2026., 2011., 2023., 1958., 1982.,
       2008., 2059., 2006., 2012., 2042., 2000., 2066., 2026., 2032.,
       2009., 2022., 1989., 2020., 2078., 2034., 2036., 1976., 2026.,
       1960., 2021., 2009., 2049., 2029., 2065., 2002., 2007., 2031.,
       2015., 2044., 2012., 1999., 2002., 1979., 1965., 2011., 2027.,
       2013., 2012., 2022., 1981., 2030., 2061., 2056., 2014., 1957.,
       1968., 2015., 2100., 1994., 2000., 1973., 1997., 1980., 2026.,
       2004., 2008., 1988., 1993., 2042., 2016., 2044., 2002., 2006.])
```

The result of this operation is that the number is added to every item in the array. Similarly,

```
bold[32, 32, 12] / 2
```

```
array([540. , 526. , 528. , 543.5, 573. , 573.5, 552.5, 532. , 564. ,
       544.5, 547.5, 524.5, 554.5, 525.5, 537. , 536.5, 556. , 543. ,
       545. , 531. , 543. , 511.5, 523.5, 569.5, 532.5, 558.5, 535. ,
       535. , 544.5, 537. , 525.5, 517. , 548. , 530. , 548. , 538. ,
       516. , 533.5, 515. , 536. , 528. , 534.5, 530.5, 527. , 536. ,
       536. , 517.5, 509. , 558. , 528. , 525.5, 542. , 537.5, 540. ,
       518. , 511. , 538. , 530. , 515.5, 539.5, 524. , 501. , 527.5,
       513.5, 507. , 503. , 536. , 501.5, 513. , 519.5, 548. , 539. ,
       512.5, 514.5, 504.5, 532.5, 511.5, 549. , 522.5, 547. , 508. ,
       507.5, 513.5, 510. , 515. , 524.5, 517. , 526.5, 509. , 519. ,
       536. , 510. , 503.5, 518.5, 541. , 525. , 505.5, 513.5, 486. ,
       496. , 492. , 510.5, 509. , 514.5, 531.5, 535. , 516.5, 524. ,
       526. , 520. , 503.5, 498. , 513. , 505.5, 511.5, 479. , 491. ,
       504. , 529.5, 503. , 506. , 521. , 500. , 533. , 513. , 516. ,
       504.5, 511. , 494.5, 510. , 539. , 517. , 518. , 488. , 513. ,
       480. , 510.5, 504.5, 524.5, 514.5, 532.5, 501. , 503.5, 515.5,
       507.5, 522. , 506. , 499.5, 501. , 489.5, 482.5, 505.5, 513.5,
       506.5, 506. , 511. , 490.5, 515. , 530.5, 528. , 507. , 478.5,
       484. , 507.5, 550. , 497. , 500. , 486.5, 498.5, 490. , 513. ,
       502. , 504. , 494. , 496.5, 521. , 508. , 522. , 501. , 503. ])
```

In this case, the number is used to separately divide each one of the elements of the array.

We can also do arithmetic between arrays. For example, we can add the values in this voxel to the items in the central voxel in the neighboring slice:

```
bold[32, 32, 12] + bold[32, 32, 11]
```

```
array([2111., 2079., 2026., 2048., 2009., 1999., 1991., 1926., 2019.,
       1940., 1964., 1969., 2015., 1979., 1942., 1976., 1993., 2016.,
       1999., 1996., 2054., 1962., 1973., 2017., 1940., 2041., 1966.,
       1960., 1957., 2000., 1966., 1961., 1974., 1965., 1953., 2033.,
       1973., 2019., 1970., 1946., 1976., 1994., 1974., 1992., 1964.,
       2017., 1961., 1946., 2015., 1959., 1959., 1999., 1945., 2031.,
       1943., 1957., 2024., 1954., 1932., 2002., 1933., 1935., 2007.,
       2012., 1917., 1931., 2032., 1953., 1976., 1951., 1983., 1966.,
       1981., 1940., 1944., 1978., 1935., 2001., 1963., 1974., 1965.,
       1939., 1997., 1930., 1959., 1952., 1950., 1981., 1982., 1938.,
       1967., 1962., 1928., 1906., 1963., 1941., 1927., 1959., 1998.,
       1971., 1891., 1975., 1938., 1985., 1939., 1940., 1914., 1939.,
       1951., 1927., 1909., 1928., 1980., 1893., 1934., 1894., 1953.,
       1901., 1989., 1881., 1928., 1959., 1914., 1962., 1908., 1948.,
       1929., 1941., 1917., 1911., 1956., 1943., 1958., 1899., 1961.,
       1917., 1933., 1922., 1920., 1911., 1944., 1908., 1916., 1969.,
       1913., 1922., 1890., 1937., 1925., 1913., 1876., 1870., 1892.,
       1957., 1932., 1930., 1906., 1882., 1913., 1946., 1939., 1947.,
       1822., 1861., 1953., 1909., 1911., 1898., 1969., 1832., 1945.,
       1878., 1952., 1885., 1918., 1931., 1892., 1913., 1950., 1911.])
```

When we perform arithmetic operations between an array and a number (or *array-scalar operations*), the number is separately added to each element. Here, when we perform arithmetic operations between arrays (*array-array operations*), the operation is instead performed element-by-element. In both cases the result is an array that has the same number of items as the original array (or arrays).

Exercises

1. Creating an array: create an array that contains the sequence of numbers from 1 to 99 in steps of two: $[1, 3, 5, \ldots, 99]$

2. Creating an array of one's: create an array of shape $(3, 5)$ containing only the number 1

3. Find all the items in an array containing numbers that have values that are larger than zero and smaller or equal to 100

4. Find all the items in an array that can be divided by three with no remainder (Hint: the function `np.mod` executes the modulo operation on every item in the array).

5. Generate an array with 100 randomly distributed numbers between 0 and 10 (Hint: look at the `np.random` submodule).

8.2.10   The SciPy Library

The NumPy array serves as the basis for numerical computing in Python. But it only does so much. Scientific computing involves a variety of different algorithms that do more interesting things with numbers; e.g., we can use fancy image processing and signal processing algorithms with our data, or we can rely on various facts about statistical distributions. Over the last 100 years, scientists and engineers have developed many different algorithms for scientific computing. In the Python scientific computing ecosystem, many of the fundamental algorithms are collected together in one library, called Scipy. It provides everything but the kitchen sink, in terms of basic algorithms for science. Because it is such a large collection of things, it is composed of several different sublibraries that perform various scientific computing tasks, such as statistics, image processing, clustering, linear algebra, and optimization. We will revisit some of these in later chapters. For now, as an example, we will focus on the signal processing sublibrary `scipy.signal`. We import the sublibrary and name it `sps`:

```
import scipy.signal as sps
```

Remember that the fMRI data set that we were looking at before had 180 time points. That means that the shape of these data were (64, 64, 25, 180). One basic signal processing operation is to efficiently and accurately resample a time series into a different sampling rate. For example, let's imagine that we would like to resample the data to a slower sampling rate. The original data were collected with a sampling interval or repetition time (TR) of 2.0 seconds. If we would like to resample this to half the sampling rate (TR = 4.0 seconds), we would halve the number of samples on the last dimension using the `sps.resample` function:

```
resampled_bold = sps.resample(bold, 90, axis=3)
print(resampled_bold.shape)
```

```
(64, 64, 25, 90)
```

If, instead, we'd like to resample this to twice the sampling rate $(TR = 1.0)$, we would double the number of samples.

```
resampled_bold = sps.resample(bold, 360, axis=3)
print(resampled_bold.shape)
```

```
(64, 64, 25, 360)
```

Although an entire book could be written about SciPy, we will not go into much more detail here. One reason is that SciPy is *under the hood* of many of the applications that you will see later in the book. For example, the high-level libraries that we will use for image processing (in part 5 of the book) and for machine learning (in part 6) rely on implementations of fundamental computational methods in SciPy. But, unless you are building such tools from scratch, you may not need to use it directly. Nevertheless, it is a wonderful resource to know about when you do want to build something new that is not already implemented in the existing libraries.

Exercise

Use the `scipy.interpolate` sublibrary to interpolate the signal in one voxel (take the central voxel, `bold[32, 32, 12]`) to even higher resolution $(TR = 0.5$ seconds, or 720 samples) using a cubic spline. Hint: look at the documentation of the `scipy.interpolate.interp1d` function to choose the method of interpolation and find example usage.

8.3   Additional Resources

For a description of the scientific Python ecosystem, see Fernando Perez, Brian Granger, and John Hunter's paper from 2011 [Perez et al. 2010]. This paper describes in some detail how Python has grown from a general-purpose programming language into a flexible and powerful tool that covers many of the uses that scientists have through a distributed ecosystem of development. More recent papers describe the tools at the core of the ecosystem: the NumPy library is described in one recent paper [Harris et al. 2020] and the SciPy library is described in another [Virtanen et al. 2020].

9

Manipulating Tabular Data
with Pandas

As we saw in a previous chapter, many kinds of neuroscience data can be described as arrays. In particular, the analysis of many physiological signals, such as the functional magnetic resonance imaging (fMRI) blood oxygenation level dependent (BOLD) signal, really benefit from organizing data in the form of arrays. But there is another very beneficial way of organizing data. In neuroscience, we often have data sets where several different variables were recorded for each observation in the data set. For example, each observation might be one subject in a study, and for each subject, we record variables like the subject's sex, age, various psychological measures, and also summaries of physiological measurements using fMRI or other brain imaging modalities. In cases like this, it is very common to represent the data in a two-dimensional table where each row represents a different observation and each column represents a different variable. Organizing our data this way allows us to manipulate the data through queries and aggregation operations in a way that makes analysis much easier. Because this type of organization is so simple and useful, it is often called *tidy data* (a more technical name, for readers with an acquaintance with databases, is the *third normal form*, or 3NF). This is also a very general way of organizing data in many different applications, ranging from scientific research to website logs, and administrative data collected in the operations of various organizations. As you will see later in the book, it is also a natural input to further analysis, such as the machine learning methods that you will see in part 6 of the book. For this reason, tools that analyze tidy tabular data play a central role in data science across these varied applications. Here, we will focus on a popular Python library that analyzes this kind of data: Pandas. Pandas is not only the plural form of an exceedingly cute animal you can use as the mascot for your software library but also an abbreviation of panel data, which is another name for data that is stored in two-dimensional tables of this sort.

An example should help demonstrate how Pandas is used. Let's consider a diffusion MRI data set. In this data set, diffusion MRI data were collected from 76 individuals ages 6–50. Let's start with the subjects and their characteristics.

We import `pandas` the usual way. Importing it as `pd` is also an oft-used convention, just like `import numpy as np`:

```
import pandas as pd
```

Pandas knows how to take sequences of data and organize them into tables. For example, the following code creates a table that has three columns, with four values in each column.

```
my_df = pd.DataFrame({'A': ['A0', 'A1', 'A2', 'A3'],
                      'B': ['B0', 'B1', 'B2', 'B3'],
                      'C': ['C0', 'C1', 'C2', 'C3']})
```

It is easy to look at all of the data in such a small table by printing out the values in the table.

```
my_df
```

```
    A    B    C
0   A0   B0   C0
1   A1   B1   C1
2   A2   B2   C2
3   A3   B3   C3
```

The Pandas library also knows how to read data from all kinds of sources. For example, it can read data directly from a comma-separated values (CSV) file stored on our computer, or somewhere on the internet. The CSV format is a good format for storing tabular data. You can write out the values in a table, making sure each row contains the same number of values. Values within a row are separated from one another by commas (hence the name of the format). To make the data more comprehensible, it is common to add a header row. That is, the first row should contain labels (or column names) that tell us what each value in that column means. We often also have one column that uniquely identifies each one of the observations (e.g., the unique ID of each subject), which can come in handy.

Here, we will point to data that was collected as part of a study of life span changes in brain tissue properties [Yeatman et al. 2014]. The data is stored in a file that contains a first row with the labels of the variables: subjectID, Age, Gender, Handedness, IQ, IQ_Matrix, and IQ_Vocab. The first of these labels is an identifier for each subject. The second is the age, stored as an integer. The third is the gender of the subject, stored as a string value (Male or Female), the fourth is handedness (Right or Left), the fifth is the IQ (also a number) and the sixth and seventh are subscores of the total IQ score (numbers as well).

We point the Pandas read_csv function directly to a URL that stores the data (though we could also point to a CSV file stored on our computer), as part of the AFQ-Browser project.[1] We also give Pandas some instructions about our data, including

---

1. https://yeatmanlab.github.io/AFQ-Browser/

which columns we would like to use; how to interpret values that should be marked as null values (in this case, the string NaN, which stands for "not a number"); and which columns in the data we want to designate as the index (i.e., the column that uniquely identifies each row in our data set). In this case, our index is the first column (`index_col=0`), corresponding to our unique subject identifier.

All of the arguments that follow (except for the URL to the file) are optional. Also, `read_csv` has several other optional arguments we could potentially use to exert very fine-grained control over how our data file is interpreted. One benefit of this flexibility is that we are not restricted to reading only from CSV files. Despite the name of the function, `read_csv` can be used to read data from a wide range of well-structured plain-text formats. And beyond `read_csv`, Pandas also has support for many other formats, via functions like `read_excel`, `read_stata`, and so on.

```
subjects = pd.read_csv(
    "https://yeatmanlab.github.io/AFQBrowser-demo/data/subjects.csv",
    usecols=[1,2,3,4,5,6,7],
    na_values="NaN", index_col=0)
```

The variable `subjects` now holds a two-dimensional table of data. This variable is an instance of an object called a `DataFrame` (often abbreviated to DF). A `DataFrame` is similar to the NumPy array objects that you saw in the previous chapter, in that it is a structured container for data. But it differs in several key respects. For one thing, a `DataFrame` is limited to two dimensions (a NumPy array can have an arbitrary number of dimensions). For another, a `DataFrame` stores a lot more metadata, i.e., data about the data. To see what this means, we can examine some of the data as it is stored in our `subjects` variable. The `.head()` method lets us see the first few rows of the table:

```
subjects.head()
```

	Age	Gender	Handedness	IQ	IQ_Matrix	IQ_Vocab
subjectID						
subject_000	20	Male	NaN	139.0	65.0	77.0
subject_001	31	Male	NaN	129.0	58.0	74.0
subject_002	18	Female	NaN	130.0	63.0	70.0
subject_003	28	Male	Right	NaN	NaN	NaN
subject_004	29	Male	NaN	NaN	NaN	NaN

A few things to notice about this table: first, in contrast to what we saw with NumPy arrays, the DataFrame tells us about the meaning of the data: each column has a label (or name). Second, in contrast to a NumPy array, our DF is *heterogeneously typed*: it contains a mixture of different kinds of variables. `Age` is an integer variable; `IQ` is a floating point (we know this because, even though the items have integer values, they indicate the decimal); and `Gender` and `Handedness` both contain string values.

We might also notice that some variables include values of NaN (which stands for "not a number"), which are now designated as null values, i.e., values that should be ignored in calculations. This is the special value we use whenever a cell lacks a valid value (due to absent or incorrect measurement). For example, it is possible that subject_003 and subject_004 did not undergo IQ testing, so we do not know what the values for the variables IQ, IQ_Matrix, and IQ_Vocab should be in these rows.

9.1   Summarizing DataFrames

Pandas provides us with a variety of useful functions for data summarization. The .info() function tells us more precisely how much data we have and how the different columns of the data are stored. It also tells us how many non-null values are stored for every column.

```
subjects.info()
```

```
<class 'pandas.core.frame.DataFrame'>
Index: 77 entries, subject_000 to subject_076
Data columns (total 6 columns):
 #   Column      Non-Null Count  Dtype
---  ------      --------------  -----
 0   Age         77 non-null     int64
 1   Gender      76 non-null     object
 2   Handedness  66 non-null     object
 3   IQ          63 non-null     float64
 4   IQ_Matrix   63 non-null     float64
 5   IQ_Vocab    63 non-null     float64
dtypes: float64(3), int64(1), object(2)
memory usage: 4.2+ KB
```

This should mostly make sense. The variables that contain string values, i.e., Gender and Handedness, are stored as object types. This is because they contain a mixture of strings, such as Male, Female, Right, or Left and values that are considered numerical, e.g., NaNs. The subject ID column has a special interpretation as the index of the array. We will use that in just a little bit.

Another view on the data is provided by a method called describe, which summarizes the statistics of each of the columns of the matrix that has numerical values, by calculating the minimum, maximum, mean, standard deviation, and quantiles of the values in each of these columns.

```
subjects.describe()
```

	Age	IQ	IQ_Matrix	IQ_Vocab
count	77.000000	63.000000	63.000000	63.000000
mean	18.961039	122.142857	60.539683	64.015873

```
std      12.246849   12.717599   7.448372   8.125015
min       6.000000    92.000000  41.000000  36.000000
25%       9.000000   114.000000  57.000000  60.000000
50%      14.000000   122.000000  61.000000  64.000000
75%      28.000000   130.000000  64.500000  70.000000
max      50.000000   151.000000  76.000000  80.000000
```

The NaN values are ignored in this summary, but the number of non-NaN values is given in a `count` row that tells us how many values from each column were used in computing these statistics. It looks like fourteen subjects in this data did not have measurements of their IQ.

9.2   Indexing into DataFrames

We have already seen in previous chapters how indexing and slicing let us choose parts of a sequence or an array. Pandas DataFrames support a variety of different indexing and slicing operations. Let's start with the selection of observations, or rows. In contrast to NumPy arrays, we cannot select rows in a DataFrame with numerical indexing. The code `subjects[0]` will usually raise an error (the exception is if we named one of our columns "0," but that would be bad practice).

One way to select rows from the data is to use the index column of the array. When we loaded the data from a file, we asked to designate the `subjectID` column as the index column for this array. This means that we can use the values in this column to select rows using the `loc` attribute of the DataFrame object. For example:

```
subjects.loc["subject_000"]
```

```
Age                20
Gender           Male
Handedness        NaN
IQ              139.0
IQ_Matrix        65.0
IQ_Vocab         77.0
Name: subject_000, dtype: object
```

This gives us a new variable that contains only data for this subject.

Notice that `.loc` is a special kind of object attribute, which we index directly with the square brackets. This kind of attribute is called an *indexer*. This indexer is label based, which means it expects us to pass the labels (or names) of the rows and columns we want to access. But often, we do not know what those labels are, and instead only know the numerical indices of our targets. Fortunately, Pandas provides an `.iloc` indexer for that purpose:

```
subjects.iloc[0]
```

```
Age                  20
Gender             Male
Handedness          NaN
IQ                139.0
IQ_Matrix          65.0
IQ_Vocab           77.0
Name: subject_000, dtype: object
```

The previous example returns the same data as before, but now we are asking for the row at position 0, rather than the row with label subject_000.

You might find yourself wondering why we only passed a single index to .loc and .iloc, given that our table has two dimensions. The answer is that it is just shorthand. In the previous code, subjects.iloc[0] is equivalent to this:

```
subjects.iloc[0, :]
```

```
Age                  20
Gender             Male
Handedness          NaN
IQ                139.0
IQ_Matrix          65.0
IQ_Vocab           77.0
Name: subject_000, dtype: object
```

You might remember the : syntax from the NumPy section. Just as in NumPy, the colon : stands for "all values." And just like in NumPy, if we omit the second dimension, Pandas implicitly assumes we want all of the values (i.e., the whole column). And *also* just like in NumPy, we can use the slicing syntax to retrieve only some of the elements (slicing):

```
subjects.iloc[0, 2:5]
```

```
Handedness          NaN
IQ                139.0
IQ_Matrix          65.0
Name: subject_000, dtype: object
```

The .iloc and .loc indexers are powerful and prevent ambiguity, but they also require us to type a few more characters. In cases where we just want to access one of the DataFrame columns directly, Pandas provides the following helpful shorthand:

```
subjects["Age"]
```

```
subjectID
subject_000    20
subject_001    31
subject_002    18
subject_003    28
subject_004    29

               ..
subject_072    40
subject_073    50
subject_074    40
subject_075    17
subject_076    17
Name: Age, Length: 77, dtype: int64
```

```
age = subjects["Age"]
```

This assigns the values in the "Age" column to the age variable. The new variable is no longer a Pandas DataFrame; it is now a Pandas Series! A Series stores a one-dimensional series of values (it is conceptually similar to a one-dimensional NumPy array) and under the hood, every Pandas DataFrame is a collection of Series (one per column). The age Series object also retains the index from the original DataFrame.

Series behave very similarly to DataFrames, with some exceptions. One exception is that, because a Series only has one dimension, we can index them on the rows without explicitly using loc or iloc (though they still work fine). So, the following two lines both retrieve the same value.

```
age['subject_072']
```

```
40
```

```
age[74]
```

```
40
```

We will see other ways that Series objects are useful in just a bit. We can also select more than one column to include. This requires indexing with a list of column names and will create a new DataFrame

```
subjects[["Age", "IQ"]]
```

```
          Age      IQ
subjectID
subject_000    20   139.0
subject_001    31   129.0
subject_002    18   130.0
subject_003    28    NaN
subject_004    29    NaN
...           ...    ...
subject_072    40   134.0
subject_073    50    NaN
subject_074    40   122.0
subject_075    17   118.0
subject_076    17   121.0

[77 rows x 2 columns]
```

We can also combine indexing operations with `loc`, to select a particular combination of rows and columns. For example:

```
subjects.loc["subject_005":"subject_010", ["Age", "Gender"]]
```

```
          Age  Gender
subjectID
subject_005    36    Male
subject_006    39    Male
subject_007    34    Male
subject_008    24  Female
subject_009    21    Male
subject_010    29  Female
```

A cautionary note: beginners often find indexing in Pandas confusing—partly because there are different `.loc` and `.iloc` indexers, and partly because many experienced Pandas users *do not* always use these explicit indexers, and instead opt for shorthand like `subjects["Age"]`. It may take a bit of time for all these conventions to sink through, but do not worry! With a bit of experience, it quickly becomes second nature. If you do not mind typing a few extra characters, the best practice (which, admittedly, we will not always follow ourselves in this book) is to always be as explicit as possible, i.e., to use `.loc` and `.iloc`, and to always include both dimensions when indexing a `DataFrame`. So, for example, we would write `subjects["Age"]` rather than the shorthand `subjects["Age"]`.

9.3   Computing with DataFrames

Like NumPy arrays, Pandas DataFrame objects have many methods for computing on the data that is in the DataFrame. However, in contrast to arrays the dimensions of the DataFrame always mean the same thing: the columns are variables and the rows are observations. This means that some kinds of computations only make sense when done along one dimension, the rows, and not along the other. For example, it might make sense to compute the average IQ of the entire sample, but it would not make sense to average a single subject's age and IQ scores.

Let's see how this plays out in practice. For example, like the NumPy array object, DataFrame objects have a mean method. However, in contrast to the NumPy array, when mean is called on the DataFrame, it defaults to take the mean of each column separately. In contrast to NumPy arrays, different columns can also have different types. For example, the "Gender" column is full of strings, and it does not make sense to take the mean of strings, so we have to explicitly pass to the mean method an argument that tells it to only try to average the columns that contain numeric data.

```
means = subjects.mean(numeric_only=True)
print(means)
```

```
Age          18.961039
IQ          122.142857
IQ_Matrix    60.539683
IQ_Vocab     64.015873
dtype: float64
```

This operation also returns to us an object that makes sense: this is a Pandas `Series` object, with the variable names of the original `DataFrame` as its index, which means that we can extract a value for any variable straightforwardly. For example:

```
means["Age"]
```

```
18.961038961038962
```

9.3.1 Arithmetic with DataFrame Columns

As you saw in the previous example, a single DataFrame column is a Pandas Series object. We can do arithmetic with Series objects in a way that is very similar to arithmetic in NumPy arrays. That is, when we perform arithmetic between a Series object and a number, the operation is performed separately for each item in the series. For example, let's compute a standardized z-score for each subject's age, relative to the distribution of ages in the group. First, we calculate the mean and standard deviation of the ages. These are single numbers:

```
age_mean = subjects["Age"].mean()
age_std = subjects["Age"].std()
print(age_mean)
print(age_std)
```

```
18.961038961038962
12.24684874445319
```

Next, we perform array-scalar computations on the "Age" column or Series. We subtract the mean from each value and divide it by the standard deviation:

```
(subjects["Age"] - age_mean ) / age_std
```

```
subjectID
subject_000     0.084835
subject_001     0.983025
subject_002    -0.078472
subject_003     0.738064
subject_004     0.819718
                  . . .
subject_072     1.717908
subject_073     2.534445
subject_074     1.717908
subject_075    -0.160126
subject_076    -0.160126
Name: Age, Length: 77, dtype: float64
```

One thing that Pandas DataFrames allow us to do is to assign the results of an arithmetic operation on one of its columns into a new column that gets incorporated into the DataFrame. For example, we can create a new column that we will call "Age_standard" and will contain the standard scores for each subject's age:

```
subjects["Age_standard"] = (subjects["Age"] - age_mean ) / age_std

subjects.head()
```

	Age	Gender	Handedness	IQ	IQ_Matrix	IQ_Vocab	Age_standard
subjectID							
subject_000	20	Male	NaN	139.0	65.0	77.0	0.084835
subject_001	31	Male	NaN	129.0	58.0	74.0	0.983025
subject_002	18	Female	NaN	130.0	63.0	70.0	-0.078472
subject_003	28	Male	Right	NaN	NaN	NaN	0.738064
subject_004	29	Male	NaN	NaN	NaN	NaN	0.819718

You can see that Pandas has added a column to the right with the new variable and its values in every row.

Exercise

In addition to array-scalar computations, we can also perform arithmetic between columns/Series (akin to array-array computations). Compute a new column in the

DataFrame called `IQ_sub_diff`, which contains in every row the difference between the subject's `IQ_Vocab` and `IQ_Matrix` columns. What happens in the cases where one (or both) of these is a null value?

9.3.2 Selecting Data

Putting these things together, we will start using Pandas to filter the data set based on the properties of the data. To do this, we can use logical operations to find observations that fulfill certain conditions and select them. This is similar to logical indexing that we saw in NumPy arrays (in Section 8.2.8). It relies on the fact that, much as we can use a Boolean array to index into NumPy arrays, we can also use a Boolean `Series` object to index into a Pandas `DataFrame`. For example, let's say that we would like to separately analyze children under age 10 and other subjects. First, we define a column that tells us for every row in the data whether the `"Age"` variable has a value that is smaller than ten:

```
subjects["age_less_than_10"] = subjects["Age"] < 10
print(subjects["age_less_than_10"])
```

```
subjectID
subject_000    False
subject_001    False
subject_002    False
subject_003    False
subject_004    False
               ...
subject_072    False
subject_073    False
subject_074    False
subject_075    False
subject_076    False
Name: age_less_than_10, Length: 77, dtype: bool
```

This column is also a Pandas `Series` object, with the same index as the original `DataFrame` and Boolean values (`True`/`False`). To select from the original `DataFrame` we use this `Series` object to index into the `DataFrame`, providing it within the square brackets used for indexing:

```
subjects_less_than_10 = subjects[subjects["age_less_than_10"]]
subjects_less_than_10.head()
```

	Age	Gender	Handedness	IQ	IQ_Matrix	IQ_Vocab	Age_standard	\
subjectID								
subject_024	9	Male	Right	142.0	72.0	73.0	-0.813355	
subject_026	8	Male	Right	125.0	67.0	61.0	-0.895009	
subject_028	7	Male	Right	146.0	76.0	73.0	-0.976663	
subject_029	8	Female	Right	107.0	57.0	51.0	-0.895009	
subject_033	9	Male	Right	132.0	64.0	71.0	-0.813355	

```
              age_less_than_10
subjectID
subject_024                True
subject_026                True
subject_028                True
subject_029                True
subject_033                True
```

As you can see, this gives us a new `DataFrame` that has all of the columns of the original `DataFrame`, including the ones that we added through computations that we did along the way. It also retains the index column of the original array and the values that were stored in this column. But several observations were dropped along the way, namely those for which the `"Age"` variable was ten or larger. We can verify that this is the case by looking at the statistics of the remaining data:

```
subjects_less_than_10.describe()
```

```
              Age         IQ   IQ_Matrix    IQ_Vocab   Age_standard
count   25.000000   24.00000   24.000000   24.000000     25.000000
mean     8.320000  126.62500   62.541667   66.625000     -0.868880
std      0.802081   14.48181    8.607273    8.026112      0.065493
min      6.000000   92.00000   41.000000   50.000000     -1.058316
25%      8.000000  119.00000   57.750000   61.750000     -0.895009
50%      8.000000  127.50000   63.500000   68.000000     -0.895009
75%      9.000000  138.00000   68.000000   72.250000     -0.813355
max      9.000000  151.00000   76.000000   80.000000     -0.813355
```

This checks out. The maximum value for age is nine, and we also have only twenty-five remaining subjects, presumably because the other subjects in the DataFrame had ages of ten or larger.

Exercises

Series objects have a `notnull` method that returns a Boolean Series object that is `True` for the cases that are not null values (e.g., NaN). Use this method to create a new DataFrame that has only the subjects for which IQ measurements were obtained. Is there a faster way to create this DataFrame with just one function call?

9.3.3   Selecting Combinations of Groups: Using a MultiIndex

Sometimes we want to select groups made up of combinations of variables. For example, we might want to analyze the data based on a split of the data both by gender and by age. One way to do this is by changing the index of the DataFrame to be made up of more than one column. This is called a MultiIndex DataFrame. A MultiIndex gives us direct access, by indexing, to all the different groups in the data, across different kinds of splits. We start by using the `.set_index()` method of the DataFrame object, to create a new kind of index for our data set. This index uses both the gender and the age group to split the data into four groups: the two gender groups ("Male" and "Female") and within each of these are the participants aged below and above ten:

```
multi_index = subjects.set_index(["Gender", "age_less_than_10"])
multi_index.head()
```

		Age	Handedness	IQ	IQ_Matrix	IQ_Vocab	\
Gender	age_less_than_10						
Male	False	20	NaN	139.0	65.0	77.0	
	False	31	NaN	129.0	58.0	74.0	
Female	False	18	NaN	130.0	63.0	70.0	
Male	False	28	Right	NaN	NaN	NaN	
	False	29	NaN	NaN	NaN	NaN	

		Age_standard
Gender	age_less_than_10	
Male	False	0.084835
	False	0.983025
Female	False	-0.078472
Male	False	0.738064
	False	0.819718

This is curious: the two first columns are now both index columns. That means that we can select rows of the DataFrame based on these two columns. For example, we can subset the data down to only the subjects identified as males under age 10 and then take the mean of that group:

```
multi_index.loc["Male", True].mean(numeric_only=True)
```

```
Age              8.285714
IQ             125.642857
IQ_Matrix       62.071429
IQ_Vocab        66.000000
Age_standard    -0.871679
dtype: float64
```

Or we can do the same operation, but only with those subjects identified as females over age 10:

```
multi_index.loc["Female", False].mean(numeric_only=True)
```

```
Age              22.576923
IQ              117.095238
IQ_Matrix        57.619048
IQ_Vocab         61.619048
Age_standard      0.295250
dtype: float64
```

While useful, this can also become a bit cumbersome if you want to repeat this many times for every combination of age group and sex. And because dividing up a group of observations based on combinations of variables is such a common pattern in data analysis, there is a built-in way to do it. Let's look at that next.

9.3.4   Split-Apply-Combine

A recurring pattern we run into when working with tabular data is the following: we want to (1) take a data set and *split* it into subsets; (2) independently *apply* some operation to each subset; and (3) *combine* the results of all the independent applications into a new data set. This pattern has a simple, descriptive (if boring) name: split-apply-combine. Split-apply-combine is such a powerful and common strategy that Pandas implements extensive functionality designed to support it. The centerpiece is a `DataFrame` method called `.groupby()` that, as the name suggests, groups (or splits) a `DataFrame` into subsets.

For example, let's split the data by the `"Gender"` column:

```
gender_groups = subjects.groupby("Gender")
```

The output from this operation is a `DataFrameGroupBy` object. This is a special kind of object that knows how to do many of the things that regular DataFrame objects can, but also internally groups parts of the original DataFrame into distinct subsets. This means that we can perform many operations just as if we were working with a regular DataFrame, but implicitly, those operations will be applied to each subset, and not to the whole DataFrame.

For example, we can calculate the mean for each group:

```
gender_groups.mean()
```

	Age	IQ	IQ_Matrix	IQ_Vocab	Age_standard	\
Gender						
Female	18.351351	120.612903	59.419355	63.516129	-0.049783	
Male	18.743590	123.625000	61.625000	64.500000	-0.017756	

	age_less_than_10
Gender	
Female	0.297297
Male	0.358974

The output of this operation is a DataFrame that contains the summary with the original DataFrame's `"Gender"` variable as the index variable. This means that we can get the mean age for one of the gender groups through a standard DataFrame indexing operation:

```
gender_groups.mean().loc["Female", "Age"]
```

```
18.35135135135135
```

We can also call `.groupby()` with more than one column. For example, we can repeat the split of the group by age groups *and* sex:

```
gender_and_age_groups = subjects.groupby(["Gender", "age_less_than_10"])
```

As before, the resulting object is a `DataFrameGroupBy` object, and we can call the `.mean()` method on it.

```
gender_and_age_groups.mean()
```

```
                             Age         IQ  IQ_Matrix   IQ_Vocab  \
Gender age_less_than_10
Female False           22.576923  117.095238  57.619048  61.619048
       True             8.363636  128.000000  63.200000  67.500000
Male   False           24.600000  122.055556  61.277778  63.333333
       True             8.285714  125.642857  62.071429  66.000000

                       Age_standard
Gender age_less_than_10
Female False                0.295250
       True                -0.865317
Male   False                0.460442
       True                -0.871679
```

The resulting object is a MultiIndex DataFrame, but rather than worry about how to work with MultiIndexes we will just use `.iloc` to retrieve the first combination of gender and age values in the previous DataFrame.

```
gender_and_age_groups.mean().iloc[0]
```

```
Age              22.576923
IQ              117.095238
IQ_Matrix        57.619048
IQ_Vocab         61.619048
Age_standard      0.295250
Name: (Female, False), dtype: float64
```

We could also have used `.reset_index()`, and then applied Boolean operations to select specific subsets of the data, just as we did earlier in this section.

Exercise

Use any of the methods you saw above to calculate the average IQ of right-handed male subjects older than 10 years.

9.4   Joining Different Tables

Another kind of operation that Pandas has excellent support for is data joining (or merging, or combining. . .). For example, in addition to the table we have been working with so far, we also have diffusion MRI data that was collected in the same individuals. Diffusion MRI uses magnetic field gradients to make the measurement in each voxel sensitive to the directional diffusion of water in that part of the brain. This is particularly useful in the brain's white matter, where diffusion along the length of large bundles of myelinated axons is much larger than across their boundaries. This fact is used to guide computational tractography algorithms that generate estimates of the major white matter pathways that connect different parts of the brain. In addition, in each voxel, we can fit a model of diffusion that tells us something about the properties of the white matter tissue within the voxel.

In the data set that we are analyzing here, the diffusion data were analyzed to extract tissue properties along the length of twenty major white matter pathways in the brain. In this analysis method, called tractometry, each major pathway is divided into 100 nodes, and in each node, different kinds of tissue properties are sampled. For example, the fractional anisotropy (FA) is calculated in the voxels that are associated with this node. This means that for every bundle, in every subject, we have exactly 100 numbers representing FA. This data can therefore also be organized in a tabular format. Let's see what that looks like by reading this table as well. For simplicity, we focus only on some of the columns in the table (in particular, there are some other tissue properties stored in the CSV file, which you can explore on your own, by omitting the `usecols` argument)

```
nodes = pd.read_csv(
    'https://yeatmanlab.github.io/AFQBrowser-demo/data/nodes.csv',
    index_col="subjectID",
    usecols=["subjectID", "tractID", "nodeID", "fa"])
```

```
nodes.info()
```

```
<class 'pandas.core.frame.DataFrame'>
Index: 154000 entries, subject_000 to subject_076
Data columns (total 3 columns):
 #   Column  Non-Null Count   Dtype
---  ------  --------------   -----
 0   tractID  154000 non-null  object
 1   nodeID   154000 non-null  int64
 2   fa       152326 non-null  float64
```

```
dtypes: float64(1), int64(1), object(1)
memory usage: 4.7+ MB
```

```
nodes.head()
```

```
                              tractID  nodeID        fa
subjectID
subject_000  Left Thalamic Radiation       0  0.183053
subject_000  Left Thalamic Radiation       1  0.247121
subject_000  Left Thalamic Radiation       2  0.306726
subject_000  Left Thalamic Radiation       3  0.343995
subject_000  Left Thalamic Radiation       4  0.373869
```

This table is much, much larger than our subjects table (154,000 rows!). This is because for every subject and every tract, there are 100 nodes and FA is recorded for each one of these nodes.

Another thing to notice about this table is that it shares one column in common with the subjects table that we looked at before: they both have an index column called subjectID. But in this table, the index values are not unique to a particular row. Instead, each row is uniquely defined by a combination of three different columns: the subjectID, and the tractID, which identifies the white matter pathway (e.g., in the first few rows of the table "Left Thalamic Radiation") as well as a nodeID, which identifies how far along this pathway these values are extracted from. This means that for each individual subject, there are multiple rows. If we index using the subjectID value, we can see just how many:

```
nodes.loc["subject_000"].info()
```

```
<class 'pandas.core.frame.DataFrame'>
Index: 2000 entries, subject_000 to subject_000
Data columns (total 3 columns):
 #   Column   Non-Null Count  Dtype
---  ------   --------------  -----
 0   tractID  2000 non-null   object
 1   nodeID   2000 non-null   int64
 2   fa       1996 non-null   float64
dtypes: float64(1), int64(1), object(1)
memory usage: 62.5+ KB
```

This makes sense: twenty major white matter pathways were extracted in the brain of each subject and there are 100 nodes in each white matter pathway, so there are a total of 2,000 rows of data for each subject.

We can ask a lot of questions with this data. For example, we might wonder whether there are sex and age differences in the properties of the white matter. To answer these kinds of questions, we need to somehow merge the information that is currently contained in two different tables. There are a few ways to do this, and some are simpler than others. Let's start by considering the simplest case, using artificial data, and then we will come back to our real data set.

Consider the following three tables, created the way we saw at the very beginning of this chapter:

```
df1 = pd.DataFrame({'A': ['A0', 'A1', 'A2', 'A3'],
                    'B': ['B0', 'B1', 'B2', 'B3'],
                    'C': ['C0', 'C1', 'C2', 'C3'],
                    'D': ['D0', 'D1', 'D2', 'D3']},
                    index=[0, 1, 2, 3])

df2 = pd.DataFrame({'A': ['A4', 'A5', 'A6', 'A7'],
                    'B': ['B4', 'B5', 'B6', 'B7'],
                    'C': ['C4', 'C5', 'C6', 'C7'],
                    'D': ['D4', 'D5', 'D6', 'D7']},
                    index=[4, 5, 6, 7])


df3 = pd.DataFrame({'A': ['A8', 'A9', 'A10', 'A11'],
                    'B': ['B8', 'B9', 'B10', 'B11'],
                    'C': ['C8', 'C9', 'C10', 'C11'],
                    'D': ['D8', 'D9', 'D10', 'D11']},
                    index=[8, 9, 10, 11])
```

Each one of these tables has the same columns: $'A'$, $'B'$, $'C'$, and $'D'$ and they each have their own distinct set of index values. This kind of data could arise if the same kinds of measurements were repeated over time. For example, we might measure the same air quality variables every week, and then store each week's data in a separate table, with the dates of each measurement as the index.

One way to merge the data from such tables is using the Pandas `concat` function (an abbreviation of *concatenation*, which is the chaining together of different elements).

```
frames = [df1, df2, df3]

result = pd.concat(frames)
```

The `result` table here would have the information that was originally stored in the three different tables, organized in the order of their concatenation.

```
result
```

	A	B	C	D
0	A0	B0	C0	D0
1	A1	B1	C1	D1
2	A2	B2	C2	D2
3	A3	B3	C3	D3
4	A4	B4	C4	D4

```
 5    A5    B5    C5    D5
 6    A6    B6    C6    D6
 7    A7    B7    C7    D7
 8    A8    B8    C8    D8
 9    A9    B9    C9    D9
10   A10   B10   C10   D10
11   A11   B11   C11   D11
```

In this case, all of our individual DataFrames have the same columns, but different row indexes, so it is very natural to concatenate them along the row axis, as shown previously. But suppose we want to merge df1 with the following new DataFrame, df4:

```
df4 = pd.DataFrame({'B': ['B2', 'B3', 'B6', 'B7'],
                    'D': ['D2', 'D4', 'D6', 'D7'],
                    'F': ['F2', 'F3', 'F6', 'F7']},
                   index=[2, 3, 6, 7])
```

Now, df4 has index values 2 and 3 in common with df1. It also has columns 'B' and 'D' in common with df1. But it also has some new index values (6 and 7) and a new column ('F') that did not exist in df1. That means that there is more than one way to put together the data from df1 and df4. The safest thing to do would be to preserve as much of the data as possible, and that is what the concat function does per default:

```
pd.concat([df1, df4])
```

```
      A    B    C    D    F
0    A0   B0   C0   D0  NaN
1    A1   B1   C1   D1  NaN
2    A2   B2   C2   D2  NaN
3    A3   B3   C3   D3  NaN
2   NaN   B2  NaN   D2   F2
3   NaN   B3  NaN   D4   F3
6   NaN   B6  NaN   D6   F6
7   NaN   B7  NaN   D7   F7
```

In this case, the new merged table contains all of the rows in df1 followed by all of the rows in df4. The columns of the new table are a combination of the columns of the two tables. Wherever possible, the new table will merge the columns into one column. This is true for columns that exist in the two DataFrame objects. For example, column B and column D exist in both of the inputs. In cases where this is not possible (because the column does not exist in one of the inputs), Pandas preserves the input values and adds NaNs, which stand for missing values, into the table. For example, df1 did not have a column called F, so the first row in the resulting table has a NaN in that column. Similarly, df4 did not have an 'A' column, so the fifth row in the table has a NaN for that column.

9.4.1 Merging Tables

There are other ways we could combine data with concat; for example, we could try concatenating only along the column axis (i.e., stacking DataFrames along their width, to create an increasingly wide result). You can experiment with that by passing axis=1 to concat and seeing what happens. But rather than belabor the point here, let's come back to our real DataFrames, which imply a more complicated merging scenario.

You might think we could just use concat again to combine our subjects and nodes data sets. But there are some complications. For one thing, our two DataFrames contain no common columns. Naively concatenating them along the row axis, as in our first concat example, would produce a rather strange looking result (feel free to try it out). But concatenating along the column axis is also tricky because we have multiple rows for each subject in nodes, but only one row per subject in subjects. It turns out that concat does not know how to deal with this type of data, and would give us an error. Instead, we need to use a more complex merging strategy that allows us to specify exactly *how* we want the merge to proceed. Fortunately, we can use the Pandas merge function for this. You see, merge is smarter than concat; it implements several standard joining algorithms commonly used in computer databases (you might have heard of inner, outer, or left joins; these are the kinds of things we are talking about). When we call merge, we pass in the two DataFrames we want to merge, and then a specification that controls how the merging takes place. In our case, we indicate that we want to merge on the index for both the left and right DataFrames (hence left_index=True and right_index=True). But we could also join on one or more columns in the DataFrames, in which case we would have passed named columns to the on keyword argument (or left_on and right_on columns).

```
joined = pd.merge(nodes, subjects, left_index=True, right_index=True)
joined.info()
```

```
<class 'pandas.core.frame.DataFrame'>
Index: 154000 entries, subject_000 to subject_076
Data columns (total 11 columns):
 #   Column          Non-Null Count   Dtype
---  ------          --------------   -----
 0   tractID         154000 non-null  object
 1   nodeID          154000 non-null  int64
 2   fa              152326 non-null  float64
 3   Age             154000 non-null  int64
 4   Gender          152000 non-null  object
 5   Handedness      132000 non-null  object
 6   IQ              126000 non-null  float64
 7   IQ_Matrix       126000 non-null  float64
 8   IQ_Vocab        126000 non-null  float64
 9   Age_standard    154000 non-null  float64
 10  age_less_than_10 154000 non-null  bool
dtypes: bool(1), float64(5), int64(2), object(3)
memory usage: 13.1+ MB
```

The result, as shown here, is a table, where each row corresponds to one of the rows in the original `nodes` table, but adds in the values from the `subjects` table that belong to that subject. This means that we can use these variables to split up the diffusion MRI data and analyze it separately by (for example) age, as before, using the split-apply-combine pattern. Here, we define our subgroups based on unique combinations of age (less than or greater than ten), tractID, and nodeID.

```
age_groups = joined.groupby(["age_less_than_10", "tractID", "nodeID"])
```

Now applying the `mean` method of the `DataFrameGroupBy` object will create a separate average for each one of the nodes in each white matter pathway, identified through the "`tractID`" and "`nodeID`" values, across all the subjects in each group.

```
group_means = age_groups.mean()
group_means.info()
```

```
<class 'pandas.core.frame.DataFrame'>
MultiIndex: 4000 entries, (False, 'Callosum Forceps Major', 0) to
  (True, 'Right Uncinate', 99)
Data columns (total 6 columns):
 #   Column         Non-Null Count  Dtype
---  ------         --------------  -----
 0   fa             4000 non-null   float64
 1   Age            4000 non-null   float64
 2   IQ             4000 non-null   float64
 3   IQ_Matrix      4000 non-null   float64
 4   IQ_Vocab       4000 non-null   float64
 5   Age_standard   4000 non-null   float64
dtypes: float64(6)
memory usage: 200.3+ KB
```

For example, the first row shows us the mean value for each variable across all subjects, for the node with ID 0, in the `Callosum Forceps Major` tract. There are 4,000 rows in all, reflecting two age groups times twenty tracts times 100 nodes. Let's select just one of the white matter pathways, and assign each age group's rows within that pathway to a separate variable. This gives us two variables, each with 100 rows (one row per node). The index for this DataFrame is a `MultiIndex`, which is why we pass tuples like (`False`, "`Left Cingulum Cingulate`") to index into it.

```
below_10_means = group_means.loc[(False, "Left Cingulum Cingulate")]
above_10_means = group_means.loc[(True, "Left Cingulum Cingulate")]
```

We can then select just one of the columns in the table for further inspection. Let's use "`fa`" as an example. To visualize the 100 numbers in this series, we will use the Matplotlib library's `plot` function (in the next chapter you will learn much more about data visualization with Matplotlib). For an interesting comparison, we can also do all of these operations but selecting the other age group as well:

```
import matplotlib.pyplot as plt

fig, ax = plt.subplots()
ax.plot(below_10_means["fa"])
ax.plot(above_10_means["fa"])
ax.set_xlabel("Node")
label = ax.set_ylabel("Fractional anisotropy")
```

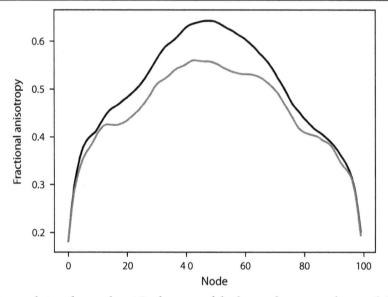

This result is rather striking! In this part of the brain, there is a substantial difference between younger children and all of the other subjects. This is a part of the white matter where there is a significant amount of development after age 10.

Exercises

How would you go about comparing the development of male and female subjects in this part of the brain? How would you compare younger children to the other subjects in other tracts?

To summarize, Pandas gives us a set of functionality to combine, query, and summarize data stored in tabular format. You have already seen here how you get from data stored in tables to a real scientific result. We will come back to using Pandas later in the book in the context of data analysis, and you will see more elaborate examples in which data can be selected using Pandas and then submitted to further analysis in other tools.

9.4.2   Pandas Errors

Before we move on to the next topic, we would like to pause and discuss a few patterns of errors that are unique to Pandas and are common enough in the daily use of

Pandas that they are worth warning you about. One common pattern of errors comes from a confusion between Series objects and DataFrame objects. These are very similar, but they are not the same thing! For example, the Series objects have a very useful `value_counts` method that creates a table with the number of observations in the Series for every unique value. However, calling this method on a DataFrame would raise a Python `AttributeError`, because it does not exist in the DataFrame, only in the Series. Another common error comes from the fact that many of the operations that you can do on DataFrames create a new DataFrame as output, rather than changing the DataFrame in place. For example, calling the following code:

```
subjects.dropna()
```

would not change the `subjects` DataFrame! If you want to retain the result of this call, you will need to assign it to a new DataFrame:

```
subjects_without_na = subjects.dropna()
```

Or use the `inplace` keyword argument:

```
subjects.dropna(inplace=True)
```

This pattern of errors is particularly pernicious because you could continue working with the unchanged DataFrame for many more steps leading to confusing results down the line.

Finally, errors due to indexing are also common. This is because, as you saw in the previous example, there are different ways to perform the same indexing operation, and, in contrast to indexing in NumPy arrays, indexing by rows and by columns, or indexing the order of a row (i.e., with `iloc`) does something rather different than indexing with the row index (i.e., with `loc`).

9.5   Additional Resources

As mentioned at the beginning of this section, Pandas is a very popular software and there are multiple examples of its use that are available online. One resource that is quite useful is a set of snippets available from Chris Albon on his website.[2] If you would like to learn more about tidy data, you can read the paper by Hadley Wickham with the same name[3] [Wickham 2014].

2. https://chrisalbon.com/
3. https://www.jstatsoft.org/index.php/jss/article/view/v059i10/v59i10.pdf

10

Visualizing Data with Python

10.1 Creating Pictures from Data

A picture is worth a thousand words. This is particularly true for large, complex, multi-dimensional data that cannot speak for itself. Visualizing data and looking at it is a very important part of (data) science. Becoming a skilled data analyst entails the ability to manipulate and extract particular bits of data, but also the ability to quickly and effectively visualize large amounts of data to uncover patterns and find discrepancies. Data visualization is also a key element in communicating about data: it is essential for reporting results compellingly and telling a coherent story. This chapter will show you how to visualize data for exploration, as well as how to create beautiful visualizations for the communication of results with others.

10.1.1 Introducing Matplotlib

There are a few different Python software libraries that visualize data. We will start with a library called Matplotlib.[1] This library was first developed almost 20 years ago by John Hunter, while he was a postdoctoral researcher in neuroscience at the University of Chicago. He needed a way to visualize time series from brain recordings that he was analyzing and, irked by the need to have a license installed on multiple different computers, he decided that instead of using the proprietary tools that were available at the time, he would develop an open-source alternative. The library grew, initially with only Hunter working on it, but eventually with many others joining him. It is now widely used by scientists and engineers in many different fields and has been used to visualize data ranging from NASA's Mars landings to Nobel-prize-winning research on gravitational waves. And of course it is still used to visualize neuroscience data. One of the nice things about Matplotlib is that it gives you a lot of control over the appearance of your visualizations, and fine-grained control over the elements of the visualization. We will get to that, but let's start with some basics.

Matplotlib is a large and complex software library, and there are several different application programming interfaces (APIs) that you can use to write your visualizations.

1. https://matplotlib.org

However, we strongly believe that there is one that you should almost always use and that gives you the most control over your visualizations. This API defines objects in Python that represent the core elements of the visualization, and allows you to activate different kinds of visualizations and manipulate their attributes through calls to methods of these objects. This API is implemented in the `pyplot` sublibrary, which we import as `plt` (another oft-used convention, similar to `import pandas as pd` and `import numpy as np`.)

```
import matplotlib.pyplot as plt
```

The first function that we will use here is `plt.subplots`. This creates the graphic elements that will constitute our data visualization.

```
fig, ax = plt.subplots()
```

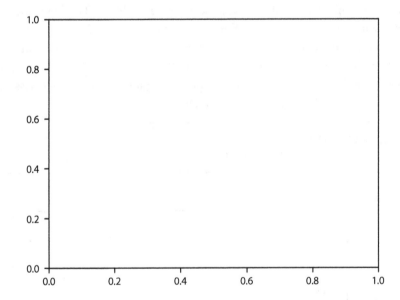

In its simplest form, the `subplots` function is called with no arguments. This returns two objects. A `Figure` object, which we named `fig`, and an `Axes` object, which we named `ax`. The `Axes` object is the element that will hold the data: you can think of it as the canvas that we will draw on with data. The `Figure` object, conversely, is more like the page onto which the `Axes` object is placed. It is a container that can hold several of the `Axes` objects (we will see that a bit later on); it can also be used to access these objects and lay them out on the screen or in a file that we save to disk.

To add data to the `Axes`, we call methods of the `Axes` object. For example, let's plot a simple sequence of data. These data are from the classic experiments of Harry Harlow's 1949 paper "The Formation of Learning Sets" [Harlow 1949]. In this experiment, animals were asked to choose between two options. In each block of trials, one of these options was rewarded with a tasty treat, while the other one was not. Here is the data from the first block of trials, a block of trials about midway through the experiment, and the last block of trials:

```
trial = [1, 2, 3, 4, 5, 6]
first_block = [50, 51.7, 58.8, 68.8, 71.9, 77.9]
middle_block = [50, 78.8, 83, 84.2, 90.1, 92.7]
last_block = [50, 96.9, 97.8, 98.1, 98.8, 98.7]
```

As you can see by reading off the numbers, the first trial in each block of the experiment had an average performance of 50%. This is because, at that point in the experiment, the animals had no way of knowing which of the two options would be rewarded. After this initial attempt, they could start learning from their experience and their performance eventually improved. But remarkably, in the first block, improvement is gradual, with the animals struggling to get to a reasonable performance on the task, while in the last session the animals improve almost immediately. In the middle blocks, improvement was neither as bad as in the initial few trials, nor as good as in the last block. This led Harlow to propose that animals are learning something about the context and to propose the (at the time) revolutionary concept of a *learning set* that animals are using in "learning to learn." Now, while this word description of the data is OK, looking at a visual of this data might be more illuminating. Let's replicate the graphic of Figure 2 from Harlow's classic paper.

In this case, we will use the `ax.plot` method:

```
fig, ax = plt.subplots()
plot = ax.plot(trial, first_block)
```

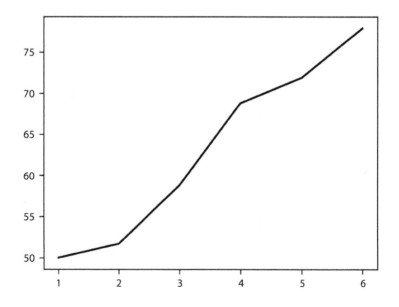

Calling `ax.plot` adds a line to the plot. The horizontal (x, or abscissa) axis of the plot represents the trials within the block, and the height of the line at each point on the vertical (y, or ordinate) dimension represents the average percent correct responses in that trial. Adding these data as a line shows the gradual learning that occurs in the first set of trials.

If you would like, you can add more data to the plot. For example, let's consider how you would add the other blocks.

```
fig, ax = plt.subplots()
ax.plot(trial, first_block)
ax.plot(trial, middle_block)
p = ax.plot(trial, last_block)
```

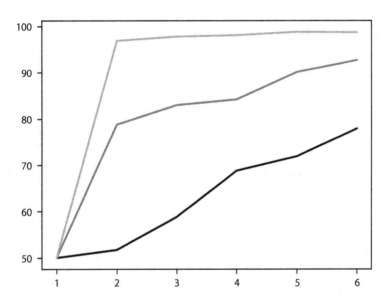

This visualization makes the phenomenon of "learning to learn" quite apparent. Something is very different about the first block of trials (black), the middle block (dark gray), and the last block (light gray). It seems as though the animal has learned that one stimulus is always going to be rewarded within a block and learns to shift to that stimulus more quickly as the experiment wears on.

But this visualization is still pretty basic and we still need to use a lot of words to explain what we are looking at. To better describe the data visually, we can add annotations to the plots. First, we would like to mark which line came from which data. This can be done by adding a label to each line as we add it to the plot and then calling `ax.legend()` to add a legend with these labels.

We would also like to add labels to the axes so that it is immediately apparent what is being represented. We can do that using the `set_xlabel` and `set_ylabel` methods of the `Axes` object.

```
fig, ax = plt.subplots()
ax.plot(trial, first_block, label="First block")
ax.plot(trial, middle_block, label="Middle block")
ax.plot(trial, last_block, label="Last block")
ax.legend()

ax.set_xlabel("Trials")
label = ax.set_ylabel("Percent correct")
```

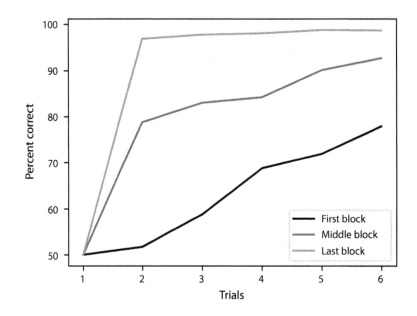

There is still more customization that we can do here before we are done. One misleading aspect of this way of showing the data is that it appears to be continuous. Because the plot function defaults to show lines with no markings indicating when the measurements were taken, it looks almost as though measurements were done between the trials in each block. We can improve this visualization by adding markers. We can choose a different marker for each one of the variables. These are added as keyword arguments to the call to plot. We will also specify a linestyle keyword argument that will modify the appearance of the lines. In this case, it seems appropriate to choose a dashed line, defined by setting linestyle='--'. Finally, using a method of the Figure object, we can also add a title.

```
fig, ax = plt.subplots()
ax.plot(trial, first_block, marker='o', linestyle='--', label="First block")
ax.plot(trial, middle_block, marker='v', linestyle='--', label="Middle block")
ax.plot(trial, last_block, marker='^', linestyle='--', label="Last block")

ax.set_xlabel("Trials")
ax.set_ylabel("Percent correct")
ax.legend()
title = ax.set_title("Harlow, 1949")
```

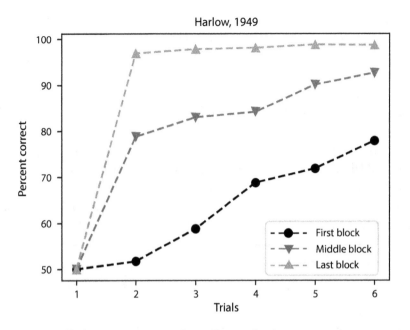

Now, this plot tells the story we wanted to tell very clearly.

Exercise

The `plot` method has multiple other keyword arguments to control the appearance of its results. For example, the `color` keyword argument controls the color of the lines. One way to specify the color of each line is by using a string that is one of the named colors specified in the Matplotlib documentation.[2] Use this keyword argument to make the three lines in the plot more distinguishable from each other using colors that you find pleasing.

2. https://matplotlib.org/stable/gallery/color/named_colors.html

---

10.1.2   Small Multiples

What if the data becomes much more complicated? For example, if there is too much to fit into just one plot? In chapter 9, we saw how to join, and then slice and dice a data set that contains measurements of diffusion MRI from the brains of seventy-seven subjects. Toward the end of chapter 9, we learned that we can visualize the data with line plots of the fractional anisotropy (FA) of different groups within the data, and we demonstrated this with just one of the twenty tracts. Now, we would like to look at several of the tracts. We grab the data using our `load_data` helper function:

```
from ndslib.data import load_data
younger_fa, older_fa, tracts = load_data("age_groups_fa")
```

Just like in chapter 9, we divided the sample into two groups: participants over the age of 10 and participants under the age of 10. Based on this division, the first two variables, `younger_fa` and `older_fa` contain Pandas DataFrame objects that hold the average diffusion fractional anistropy in each node along the length of twenty different brain white matter pathways for each one of these groups. The `tracts` variable is a Pandas `Series` object that holds the names of the different pathways (Left Corticospinal, Right Corticospinal, etc.).

One way to plot all of this data is to iterate over the names of the pathways and, in each iteration, add the data for one of the tracts into the plot. We will always use the same color for all of the curves derived from the young people; using `"C0"` selects the first of the standard colors that are used by Matplotlib. In our setup, that is black. Similarly, we use `"C1"`, a lighter gray, for the lines of FA for older subjects.

```
fig, ax = plt.subplots()
for tract in tracts:
    ax.plot(younger_fa[tract], color="C0")
    ax.plot(older_fa[tract], color="C1")
ax.set_xlabel("Node")
label = ax.set_ylabel("Fractional Anisotropy")
```

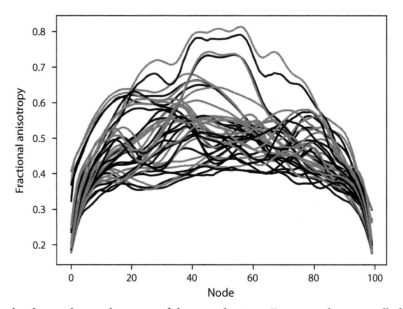

It is hard to make anything out of this visualization. For example, it is really hard to distinguish which data comes from which brain pathway and to match the two different lines—one for young subjects and the other for older subjects—that were derived from each of the pathways. This makes it hard to make out the patterns in this data. For example, it looks like maybe in some cases there is a difference between younger and older participants, but how would we tell in which pathways these differences are most pronounced?

One way to deal with this is to separate the plots so that the data from each tract has its own set of axes. We could do this by initializing a separate `Figure` object for each tract, but we can also do this by using different subplots. These are different parts of the same `Figure` object that each have their own set of axes. The `subplots` function that you saw before takes as inputs the number of subplots that you would like your figure to have. In its simplest form, we can give it an integer and it would give us a sequence of `Axes` objects, with each subplot taking up a row in the figure. Since both subplots have the same *x*-axis, we label the axis only in the bottom subplot.

```
fig, [ax1, ax2] = plt.subplots(2)
ax1.plot(younger_fa["Right Arcuate"])
ax1.plot(older_fa["Right Arcuate"])
ax2.plot(younger_fa["Left Arcuate"])
ax2.plot(older_fa["Left Arcuate"])
ax2.set_ylabel("Fractional Anisotropy")
ax1.set_ylabel("Fractional Anisotropy")
label = ax2.set_xlabel("Node")
```

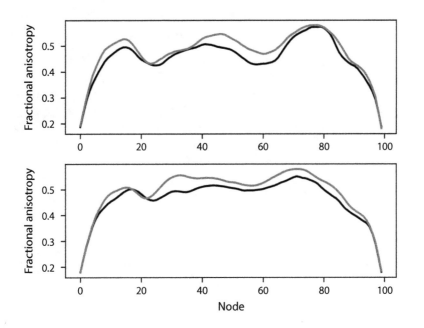

To make this a bit more elaborate, we can give `subplots` two numbers as inputs. The first number determines the number of rows of subplots and the second number determines the number of columns of plots. For example, if we want to fit all of the pathways into one figure, we can ask for five rows of subplots and in each row to have four subplots (for a total of $5 \times 4 = 20$ subplots). Then, in each subplot, we can use the data only for that particular pathway. To make things less cluttered, after adding a small title for each subplot, we will also remove the axis in each plot and use the `set_layout_tight` method of the figure object, so that each one of the subplots is comfortably laid out within the confines of the figure. Finally, to fit all this in one figure, we can also make the figure slightly larger.

```python
fig, ax = plt.subplots(5, 4)
for tract_idx in range(20):
    pathway = tracts[tract_idx]
    ax.flat[tract_idx].plot(younger_fa[pathway], color="C0")
    ax.flat[tract_idx].plot(older_fa[pathway], color="C1")
    ax.flat[tract_idx].set_title(pathway)
    ax.flat[tract_idx].axis("off")

fig.set_tight_layout("tight")
fig.set_size_inches([10, 8])
```

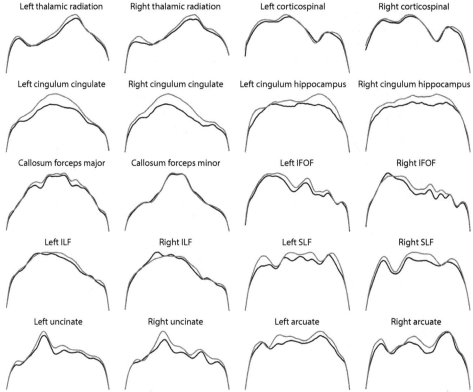

Repeating the same pattern multiple times for different data that have the same basic structure can be quite informative and is sometimes referred to as a set of *small multiples*. For example, small multiples are sometimes used to describe multiple time series of stock prices or temperatures over time in multiple different cities. Here, it teaches us that some brain pathways tend to change between the younger and older participants, while others change less. Laying the curves out in this way also shows us the symmetry of pathways in the two hemispheres. Because we have removed the axes, we cannot directly compare the values between pathways, though.

Exercise

The Axes objects have set_ylim() and set_xlim() methods, which set the limits of each of the dimensions of the plot. They take a list of two values for the minimal value of the range. To facilitate comparisons between tracts, use the set_ylim method to set the range of the FA values in each of the plots so that they are the same in all of the plots. To make sure that the code you write provides the appropriate range in all of the subplots, you will need to start by figuring out what are the maximal and minimal values of FA in the entire data set.

10.2   Scatter Plots

Another way to compare data sets uses scatter plots. A scatter plot contains points that are placed into the axes so that the position of each marker on one axis (i.e., the abscissa) corresponds to a value in one variable and the position on the other axis (i.e., the ordinate) corresponds to the value in another variable. The scatter plot lets us compare these two values to each other directly, and also allows us to see the patterns overall. For example, in our case, each point in the plot could represent one of the tracts. The position of the marker on the abscissa could be used to represent the average value of the FA in this tract in the younger subjects, while the position of this same marker on the ordinate is used to represent the average FA in this tract in the older subjects.

One way to see whether the values are systematically different is to add the $y = x$ line to the plot. Here, this is done by plotting a dashed line between $x = 0.3, y = 0.3$ and $x = 0.7$, $y = 0.7$:

```
fig, ax = plt.subplots()

ax.scatter(younger_fa.groupby("tractID").mean(),
           older_fa.groupby("tractID").mean())

ax.plot([0.3, 0.7], [0.3, 0.7], linestyle='--')
ax.set_xlabel("FA (young)")
label = ax.set_ylabel("FA (old)")
```

We learn a few different things from this visualization (which appears on the next page). First, we see that the average FA varies quite a bit between the different tracts and that the variations in young people roughly mimic the variations in older individuals. That is why points that have a higher x value, also tend to have a higher y value. In addition, there is also a systematic shift. For almost all of the tracts, the x value is slightly lower than the y value, and the markers lie above the equality line. This indicates that the average FA for the tracts is almost always higher in older individuals than in younger individuals. We saw this pattern in the small multiples earlier, but this plot demonstrates how persistent this pattern is and how consistent its size is, despite

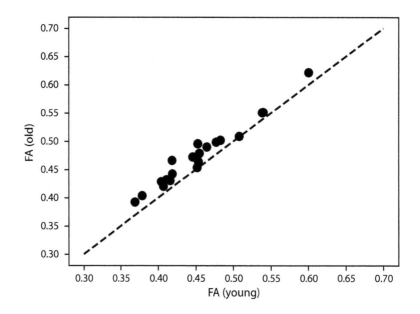

large differences in the values of FA in different tracts. This is another way to summarize and visualize the data that may make the point.

10.3   Statistical Visualizations

Statistical visualizations are used to make inferences about the data or comparisons between different data. To create statistical visualizations, we will rely on the Seaborn library.[3] This library puts together the power of Pandas with the flexibility of Matplotlib. It was created by Michael Waskom, while he was a graduate student in neuroscience at Stanford. Initially, he used it for his research purposes, but he made it publicly available and it quickly became very popular among data scientists. This is because it makes it easy to create very elegant statistical visualizations with just a bit of code.

To demonstrate what we mean by statistical visualizations, we will turn to look at the table that contains the properties of the subjects from the Pandas chapter.

```
import pandas as pd
import seaborn as sns

subjects = pd.read_csv(
  "https://yeatmanlab.github.io/AFQBrowser-demo/data/subjects.csv",
  usecols=[1,2,3,4,5,6,7],
  na_values="NaN", index_col=0)
```

First, let's draw comparisons of IQ between people with different handedness and gender. We will use a bar chart for that. This chart draws bars whose height signifies the average

3. https://seaborn.pydata.org/

value of a variable among a group of observations. The color of the bar and its position may signify the group to which this bar belongs. Here, we will use both. We will use the *x*-axis to split the data into right-handed and left-handed individuals and we will use the color of the bar (the more technically accurate term is *hue*) to split the data into male-identified and female-identified individuals:

```
b = sns.barplot(data=subjects, x="Handedness", y="IQ", hue="Gender")
```

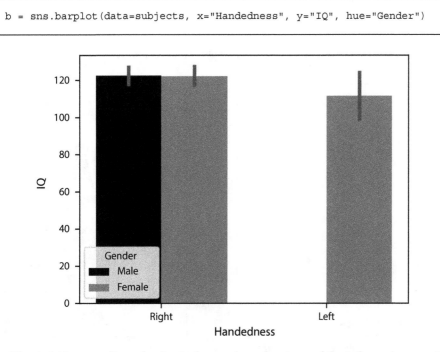

That is it! Just one line of code. Seaborn takes advantage of the information provided in the Pandas DataFrame object itself to infer how this table should be split up and how to label the axes and create the legend. In addition to drawing the average of each group as the height of the bar, Seaborn adds error bars—the vertical lines at the top of each bar—that bound the 95% confidence interval of the values in the bar. These are computed under the hood using a procedure called bootstrapping, where the data are randomly resampled multiple times to assess how variable the average is. In this case, there are no left-handed males, so that bar remains absent.

Bar charts are mostly OK, especially if you do not truncate the *y*-axis to exaggerate the size of small effects, and if the mean is further contextualized using error bars that provide information about the variability of the average quantity. But the truth is that bar charts also leave a lot to be desired. A classical demonstration of the issues with displaying only the mean and error of the data is in the so-called Anscombe's quartet. These are four sets of data that were cooked up by the statistician Francis Anscombe so that they each have the same mean, the same variance, and even the same correlation between the two variables that comprise each data set. This means that the bar charts representing each of these data sets would be identical, even while the underlying data have quite different interpretations.

There are better ways to display data. Seaborn implements several different kinds of visualizations that provide even more information about the data. It also provides a variety of options for each visualization. For example, one way to deal with the limitations of the bar chart is to show every observation within a group of observations, in what is called a *swarmplot*. The same data, including the full distribution, can be visualized as a smoothed silhouette. This is called a *violin plot*.

```
fig, ax = plt.subplots(1, 2)
g = sns.swarmplot(data=subjects, x="Handedness", y="IQ", ax=ax[0])
g = sns.violinplot(data=subjects, x="Handedness", y="IQ", ax=ax[1])
```

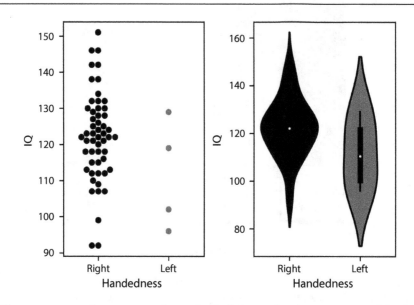

Choosing among these options depends a little bit on the characteristics of the data. For example, the swarmplot visualization shows very clearly that there is a discrepancy in the size of the samples for the two groups, with many more data points for the right-handed group than for the left-handed group. Conversely, if there is a much larger number of observations, this kind of visualization may become too crowded, even hard to read. The violin plot would potentially provide a more useful summary of the distribution of data in such a case. Admittedly, these choices are also based to some degree on personal preference.

Another way to summarize these data is the classic boxplot. This plot follows a particular convention: the median of the distribution is displayed as a vertical line. The quartiles of the distribution (i.e., the twenty-fifth and seventy-fifth percentiles) are shown as the bottom (twenty-fifth) and top (seventy-fifth) of a solid-colored bar. Then, the range of the distribution can be shown as whiskers, or horizontal lines, at the end of a vertical line extending beyond. Sometimes certain data points will appear even further beyond these whiskers. These observations are identified as outliers. This is usually done by comparing how far they are from one of the quartiles, relative to how far the quartiles are from each other. For example, in the IQ data that we have been plotting here, one right-handed subject is identified as an outlier because the difference between their score and the twenty-fifth percentile is larger than 1.5 times the difference between the score of the twenty-fifth and the seventy-fifth percentiles.

```
g = sns.boxplot(data=subjects, x="Handedness", y="IQ")
```

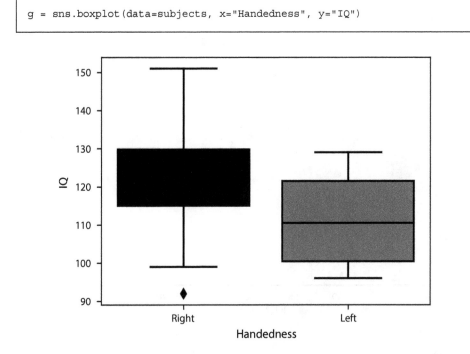

One advantage of the boxplot is that readers familiar with the conventions guiding its design will be able to glean a wealth of information about the data simply by glancing at the rather economical set of markings in this plot. Conversely, as with the violin plot, it does not communicate some aspects of the data, e.g., the discrepancy in sample size between the left-handed and right-handed sample.

Statistical visualization can accompany even more elaborate statistical analysis than just the distribution of the data. For example, Seaborn includes functionality to fit a linear model to the data. This is done using the `lmplot` function. This function visualizes not only the scatter plot of the data (in the following example, the scores in two subtests of the IQ test), but also a fit of a linear regression line between the variables, and a shaded area that signifies that 95% confidence interval derived from the model. In the example to follow, we show that this functionality can be invoked to split up the data based on another column in the data; e.g., the gender column (figure appears on the subsequent page).

```
g = sns.lmplot(data=subjects, x="IQ_Matrix", y="IQ_Vocab", hue="Gender")
```

One of the strengths of Seaborn is that it produces images that are both aesthetically pleasing and highly communicative. In most data sets, it can produce the kinds of figures that you would include in a report or scientific paper.

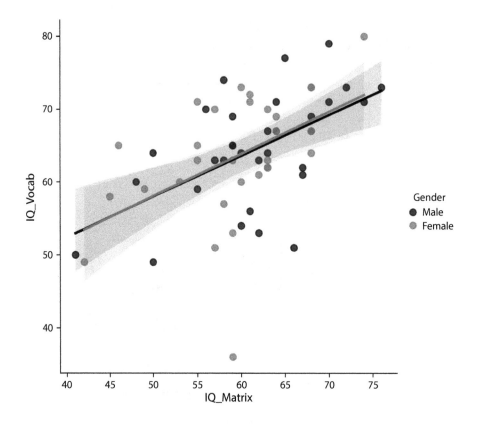

10.3.1   Visualizing and Exploring Image Data

Visualizations are a very effective tool for communicating about data, but another one of the purposes of data visualization is to aid in the process of research. For example, it is always a good idea to spend time looking at your raw and processed data. This is because statistical summaries of data can be misleading. For example, a single corrupted image with values much higher or much lower than all of the other images in an MRI series may drive the mean signal up or down substantially. And there are other ways in which you may be misled if you do not take the time to look at the data closely. This means that it is good to have at your fingertips different ways in which you can quickly visualize some data and explore it. For example, we can use Matplotlib to directly visualize a single slice from the mean of a blood oxygenation level dependent (BOLD) functional magnetic resonance imaging fMRI time series:

```
import numpy as np
from ndslib import load_data
bold = load_data("bold_numpy", fname="bold.npy")
bold.shape
fig, ax = plt.subplots()
im = ax.matshow(np.mean(bold, -1)[:, :, 10], cmap="bone")
```

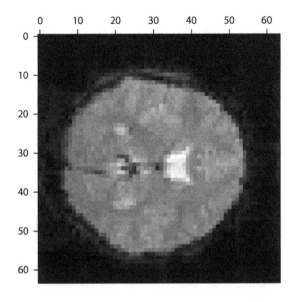

Exercise

Use the code shown previously to explore the fMRI time series in more detail. Can you create a figure where the tenth slice in every time point is displayed as part of a series of small multiples? How would you make sure that each image shows the same range of pixel values (Hint: explore the keyword arguments of matshow).

10.4   Additional Resources

The Matplotlib documentation is a treasure trove of ideas about how to visualize your data. In particular, there is an extensive gallery[4] of fully worked-out examples that show how to use the library to create a myriad of different types of visualizations.

John Hunter, who created Matplotlib, died at a young age from cancer. The scientific Python community established an annual data visualization contest in his honor that is held each year as part of the SciPy conference. The examples on the contest website[5] are truly inspiring in terms of the sophistication and creative use of data visualizations.

To learn more about best practices in data visualization, one of the definitive sources is the books of Edward Tufte. As we mentioned previously, choosing how to visualize your data may depend to some degree on your personal preferences. But there is value in receiving authoritative guidance from experts who have studied how best to communicate with data visualization. Tufte is one of the foremost authorities on the topic. And though his books tend to be somewhat curmudgeonly, they are an invaluable source of ideas and

---

4. https://matplotlib.org/stable/gallery/index.html
5. https://jhepc.github.io/

opinions about visualization. To get you started, we would recommend his first book (of at least five!) on the topic, *The Visual Display of Quantitative Information*.

If you are interested in another way to show all of the points in a data set, you can also look into so-called *raincloud plots*,[6] which combine swarmplots, box plots, and silhouette plots all in one elegant software package.

There are many other libraries for data visualization in Python. The pyviz website[7] provides an authoritative, community-managed resource, which includes an extensive list of visualization tools in Python, organized by topics and features, and links to tutorials for a variety of tools.

6. https://github.com/RainCloudPlots/RainCloudPlots
7. https://pyviz.org/

Neuroimaging in Python

11

Data Science Tools for Neuroimaging

In previous chapters, we saw that there are plenty of things that you can do with general purpose software and data science tools. Many of the examples used data from neuroimaging experiments, but the ideas that we introduced—arrays and computing with arrays, tables and how to manipulate tabular data, scientific computing, and data visualization—were not specific to neuroimaging. In this chapter, we will discuss data science approaches that are specifically tailored to neuroimaging data. First (in Section 11.1), we will present a survey of neuroimaging-specific software implemented in Python. Next (in Section 11.2), we will discuss how to organize data from neuroscience experiments for data analysis that best take advantage of these software tools. In the following chapters, we will dive into some of the applications of these data science tools to fundamental problems in neuroimaging data analysis.

11.1 Neuroimaging in Python

Within the broader ecosystem of Python tools for science, there is a family of tools specifically focused on neuroimaging (we will refer to them collectively as NiPy, which stands for Neuroimaging in Python). These software tools, developed by and for neuroimaging researchers, provide a wide range of data analysis tasks that can be used on a variety of experimental data. In the next few sections, we will see in detail how some of these tools are used. But first, we will provide a broad survey of the different kinds of tools that currently exist. It is important to emphasize that this is a very dynamically evolving ecosystem, and some of these tools may evolve over time, or even disappear. New ones will inevitably also emerge. So this survey will be, by necessity, a bit superficial and dated. That said, we will try to give you a sense of how an ecosystem like this one emerges and evolves so that you can keep an eye on these trends as they play out in the future.

11.1.1 Reading and Writing Neuroimaging Data

All of the NiPy tools depend to some degree on the ability to read neuroimaging data from commonly used file formats into the memory of the computer. The tool of choice for doing that is called NiBabel. The name hints at the plethora of different file formats that exist to represent and store neuroimaging data. NiBabel, which has been in continuous development since 2007, provides a relatively straightforward common route to read many of these formats. Much in the way that NumPy serves as a basis for all of the scientific computing in Python, NiBabel serves as the basis for the NiPy ecosystem. Because it is so fundamental, we devote an entire chapter (chapter 12) to some basic operations that you can do with NiBabel. In chapter 13, we will go even deeper to see how NiBabel enables the integration of different neuroimaging data types; as a result, its mention in this chapter will be brief.

11.1.2   Tools for MRI Data Analysis

Tools for neuroimaging data analysis are usually focused on particular data types and on particular ways of analyzing the data. Functional MRI (fMRI) is a very popular data type in cognitive neuroscience, and several tools focus on fMRI. For example, both the BrainIAK[1] software library and the Nilearn[2] software library focus on machine learning analysis of fMRI data. We will return to talk about machine learning in more detail in part 6 of the book. Some tools fit more traditional generalized linear models (GLM) to fMRI data. One of these is fitlins.[3] It relies heavily on the Brain Imaging Data Structure (BIDS) standard, which we will discuss subsequently. Another one is glmsingle,[4] implemented in both MATLAB and Python, which focuses on the estimation of single-trial responses.

Another popular MRI modality is diffusion MRI (dMRI). It is used to characterize the properties of brain white matter and to find white matter connections between different parts of the brain. Diffusion Imaging in Python[5] (DIPY) is a software library that implements a variety of methods for analysis of this kind of data. We will use image processing algorithms that are implemented in DIPY in chapter 16. Building upon DIPY, there are several tools that focus on more specific dMRI analysis tasks, such as DMIPY,[6] which focuses on various ways to model the microstructure of brain white matter tissue, pyAFQ[7] and TractSeg,[8] which focus on delineating the major white matter tracts, and NDMG[9] which focuses on quantifying the connectivity matrix of the brain based on dMRI data. Another widely used software library for analysis of dMRI is the Medical Imaging Toolkit for Diffusion.[10]

Other tools focus on analysis of structural MRI data. The advanced normalization tools (ANTs) software library implements a variety of algorithms for structural data, such as

1. https://brainiak.org/
2. https://nilearn.github.io/
3. https://fitlins.readthedocs.io/
4. https://glmsingle.readthedocs.io/
5. https://dipy.org
6. https://dmipy.readthedocs.io/en/latest/
7. http://yeatmanlab.github.io/pyAFQ/
8. https://github.com/MIC-DKFZ/TractSeg
9. https://ndmg.neurodata.io/
10. https://github.com/MIC-DKFZ/MITK-Diffusion

image registration and image segmentation, which will be described in more detail in chapter part 5 of the book. The ANTs library, originally developed in the C programming language, is now also available through a user-friendly Python interface in the antspy library.[11]

11.1.3 Tools for Other Neuroimaging Modalities

One of the largest and most mature projects for neuroimaging data analysis focuses on electrophysiology data that comes from experiments that use magnetoencephalography (MEG), electroencephalography (EEG), electrocorticography (ECoG), and other similar measurements. The MNE[12] software library has been in development since 2011 and contains implementations of many algorithms that are common across these different types of data, e.g., analysis of the frequency spectrums of these signals. It also includes algorithms that are specific for particular types of data, such as algorithms to localize the brain sources of MEG signals that are measured outside the skull.

11.1.4 Nipype, the Python Tool to Unify Them All

One of the main challenges for anyone doing neuroimaging data analysis is that different algorithms are implemented in different software libraries. Even if you decide that you would like to use NiPy tools for most of your data analysis, you might still find yourself needing to use non-Python tools for some steps in your analysis. Some of these might include novel algorithms that are implemented in the popular FMRIB Software Library (FSL) or Analysis of Functional NeuroImages (AFNI) frameworks for fMRI, the popular MRTRIX library for dMRI, or FreeSurfer for analysis of structural MRI. In theory, if you wanted to construct an analysis pipeline that glues together operations from different frameworks, you would have to familiarize yourself with the various ways frameworks expect their input data to appear and the myriad ways they produce their output. The complexity can quickly become daunting. Moreover, creating reproducible data analysis programs that put together such pipelines becomes cumbersome and can be very difficult. Fortunately, these challenges have been addressed through the Nipype[13] Python software library. Nipype implements Python interfaces to many neuroimaging software tools (including those we mentioned previously). One of Nipype's appealing features is that it provides a very similar programming interface across these different tools. In addition, it provides the functionality to represent pipelines of operations, where the outputs of one processing step become the inputs to the next processing steps. Running such a pipeline as one program is not only a matter of convenience but also important for reproducibility. By tracking and recording the set of operations that a data set undergoes through analysis, as well as the parameters that are used in each step, Nipype allows users of the software to

11. https://antspy.readthedocs.io/
12. https://mne.tools/
13. https://nipype.readthedocs.io/

report exactly what they did with their data to come to a particular conclusion. In addition, Nipype allows a processing pipeline to be used with other data sets. However, this raises another kind of challenge: to achieve this kind of extensibility, the input data have to be organized in a particular way that the software recognizes. This is where community-developed data standards come in. In the rest of this chapter, we will discuss BIDS, which is the most commonly used data standard for organizing neuroimaging data.

11.2   The Brain Imaging Data Structure Standard

In the previous section, we pointed out that Nipype can be used to create reproducible analysis pipelines that can be applied across different data sets. This is true, in principle, but in practice it also relies on one more idea: a data standard. This is because to be truly transferable, an analysis pipeline needs to know where to find the data and metadata that it uses for analysis. Thus, in this section, we will shift our focus to talking about how entire neuroimaging *projects* are (or should be) laid out. Until recently, data sets were usually organized in an idiosyncratic way. Each researcher had to decide on their own what data organization made sense to them. This made data sharing and reuse quite difficult because if you wanted to use someone else's data, there was a very good chance you would first have to spend a few days just figuring out what you were looking at and how you could go about reading the data into whatever environment you were comfortable with.

11.2.1   The BIDS Specification

Fortunately, things have improved dramatically in recent years. Recognizing that working with neuroimaging data would be considerably easier if everyone adopted a common data representation standard, a group of (mostly) fMRI researchers convened in 2015 to create something now known as BIDS.[14] BIDS was not the *first* data standard proposed in fMRI, but it has become by far the most widely adopted. Much of the success of BIDS can be traced to its simplicity: the standard deliberately insists not only on machine readability but also on human readability, which means that a machine can ingest a data set and do all kinds of machine processing with it, but a human looking at the files can also make sense of the data set, understanding what kinds of data were collected and what experiments were conducted. While there are some nuances and complexities to BIDS, the core of the specification consists of a relatively simple set of rules a human with some domain knowledge can readily understand and implement.

We will not spend much time describing the details of the BIDS specification in this book, as there is already excellent documentation for that on the project's website. Instead, we will just touch on a couple of core principles. The easiest way to understand what BIDS is about is to dive right into an example. Here is a sample directory structure we have

14. https://bids.neuroimaging.io

borrowed from the BIDS documentation. It shows a valid BIDS data set that contains just a single subject.

```
project/
    sub-control01/
        anat/
            sub-control01_T1w.nii.gz
            sub-control01_T1w.json
            sub-control01_T2w.nii.gz
            sub-control01_T2w.json
        func/
            sub-control01_task-nback_bold.nii.gz
            sub-control01_task-nback_bold.json
            sub-control01_task-nback_events.tsv
            sub-control01_task-nback_physio.tsv.gz
            sub-control01_task-nback_physio.json
            sub-control01_task-nback_sbref.nii.gz
        dwi/
            sub-control01_dwi.nii.gz
            sub-control01_dwi.bval
            sub-control01_dwi.bvec
        fmap/
            sub-control01_phasediff.nii.gz
            sub-control01_phasediff.json
            sub-control01_magnitude1.nii.gz
            sub-control01_scans.tsv
    code/
        deface.py
    derivatives/
    README
    participants.tsv
    dataset_description.json
    CHANGES
```

There are two important points to note here. First, the BIDS specification imposes restrictions on how files are organized within a BIDS project directory. For example, every subject's data go inside a sub-[id] folder below the project root where the sub- prefix is required, and the [id] is a researcher-selected string uniquely identifying that subject within the project ("control01" in the example). And similarly, inside each subject directory, we find subdirectories containing data of different modalities: anat for anatomical images; func for functional images; dwi for diffusion-weighted images; and so on. When there are multiple data collection sessions for each subject, an extra level is introduced to the hierarchy, so that functional data from the first session acquired from subject control01 would be stored inside a folder like sub-control01/ses-01/func.

Second, valid BIDS files must follow particular naming conventions. The precise naming structure of each file depends on what kind of file it is, but the central idea is that a BIDS filename is always made up of (1) a sequence of key-value pairs, where each key is separated from its corresponding value by a dash, and pairs are separated by underscores; (2) a suffix that directly precedes the file extension and describes the type of data contained in the file (this comes from a controlled vocabulary, meaning that it can only be one of a few accepted values, such as `"bold"` or `"dwi"`); and (3) an extension that defines the file format.

For example, if we take a file like `sub-control01/func/sub-control01_task-nback_bold.nii.gz` and examine its constituent chunks, we can infer from the filename that the file is a NIfTI image (`.nii.gz` extension) that contains blood oxygenation level dependent (BOLD) functional MRI (fMRI) data (`bold` suffix) for task `nback` acquired from subject `control01`.

Besides these conventions, there are several other key elements of the BIDS specification. We will not discuss them in detail, but it is good to at least be aware of them:

- Every data file should be accompanied by a JavaScript Object Notation (JSON) sidecar containing metadata describing that file. For example, a BOLD data file might be accompanied by a sidecar file that describes acquisition parameters, such as repetition time.
- BIDS follows an *inheritance* principle, meaning that JSON metadata files higher up in the hierarchy automatically apply to relevant files lower in the hierarchy unless explicitly overridden. For example, if all of the BOLD data in a single data set was acquired using the same protocol, this metadata need not be replicated in each subject's data folder.
- Every project is required to have a `dataset_description.json` file at the root level that contains basic information about the project (e.g., the name of the data set and a description of its constituents, as well as citation information).
- BIDS does not actively prohibit you from including non-BIDS-compliant files in a BIDS project, so you do not have to just throw out files that you cannot easily shoehorn into the BIDS format. The downside of including noncompliant files is just that most BIDS tools and/or human users will not know what to do with them, so your data set might not be quite as useful as it otherwise would be.

BIDS DERIVATIVES

The BIDS specification was originally created with static representations of neuroimaging data sets in mind. But it quickly became clear that it would also be beneficial for the standard to handle *derivatives* of data sets, that is, new BIDS data sets generated by applying some transformation to one or more existing BIDS data sets. For example, let's suppose we have a BIDS data set containing raw fMRI images. Typically, we will want to preprocess our images (e.g., to remove artifacts, apply motion correction, temporally filter the signal) before submitting them to analysis. It is great if our preprocessing pipeline can take BIDS data sets as inputs, but what should it then do with the output? A naive approach would be to just construct a new BIDS data set that is very similar to the original one, but replace the original (raw) fMRI images with new (preprocessed) ones. But that is likely to confuse: a user could easily end up with many different versions of the same BIDS data set, yet have no formal way to determine the relationship between them. To address this problem, the BIDS-Derivatives extension introduces some additional metadata and file naming conventions that make it easier to chain BIDS-aware tools (described in the next section) without chaos taking hold.

11.2.2   The BIDS Ecosystem

At this point, you might be wondering: *what is BIDS good for?* Surely the point of introducing a new data standard is not just to inconvenience people by forcing them to spend their time organizing their data a certain way? There must be some benefits to individual researchers—and ideally, the community as a whole—spending precious time making data sets and workflows BIDS compliant, right? Well, yes, there are! The benefits of buying into the BIDS ecosystem are quite substantial. Let's look at a few.

EASIER DATA SHARING AND REUSE

One obvious benefit we alluded to earlier is that sharing and reusing neuroimaging data become much easier once many people agree to organize their data the same way. As a trivial example, once you know that BIDS organizes data according to a fixed hierarchy (i.e., subject -> session -> run), it is easy to understand other people's data sets. There is no chance of finding time-course images belonging to subject 1 in, say, `/imaging/old/NbackTask/rawData/niftis/controlgroup/1/`. But the benefits of BIDS for sharing and reuse come into full view once we consider the impact on public data repositories. While neuroimaging repositories have been around for a long time (for an early review, see Van Horn et al. [2001]), their utility was long hampered by the prevalence of idiosyncratic file formats and project organizations. By supporting the BIDS standard, data repositories open the door to a wide range of powerful capabilities.

To illustrate, consider OpenNeuro[15], which is currently the largest and most widely used repository of brain MRI data. OpenNeuro requires uploaded data sets to be in BIDS format (though data sets do not have to be fully compliant). As a result, the platform can automatically extract, display, and index important metadata. This can include the number of subjects, sessions, and runs in each data set; the data modalities and experimental tasks present; a standardized description of the data set; and so on. Integration with free analysis platforms like BrainLife[16] is possible, as is structured querying over datasets via OpenNeuro's GraphQL application programming interface (API) endpoint.

Perhaps most importantly, the incremental effort required by users to make their BIDS-compliant data sets publicly available and immediately usable by others is minimal: in

15. https://openneuro.org
16. https://brainlilfe.io

most cases, users have only to click an Upload button and locate the project they wish to share (there is also a command-line interface, for users who prefer to interact with OpenNeuro programmatically).

BIDS APPS

A second benefit to representing neuroimaging data in BIDS is that one immediately gains access to a large, and rapidly growing ecosystem of BIDS-compatible tools. If you have used different neuroimaging tools in your research—for example, perhaps you have tried out both FSL and SPM (the two most widely used fMRI data analysis suites)—you will probably have done some work to get your data into a somewhat different format for each tool. In a world without standards, tool developers cannot be reasonably expected to know how to read *your* particular data set, so the onus falls on *you* to get your data into a compatible format. In the worst case, this means that every time you want to use a new tool, you have to do some more work.

By contrast, for tools that natively support BIDS, life is simpler. Once we know that fMRIPrep, which is a very popular preprocessing pipeline for fMRI data [Esteban et al. 2019], takes valid BIDS data sets as inputs, the odds are very high that we will be able to apply fMRIPrep to our own valid BIDS data sets with little or no additional work. To facilitate the development and use of these kinds of tools, BIDS developed a lightweight standard for BIDS Apps [Gorgolewski et al. 2017]. A BIDS App is an application that takes one or more BIDS data sets as input. There is no restriction on what a BIDS App can do, or what it is allowed to output (though many BIDS Apps output BIDS-Derivatives data sets); the only requirement is that a BIDS App is containerized (using Docker or Singularity; see chapter 4), and accepts valid BIDS data sets as input. New BIDS Apps are continuously being developed, and as of this writing, the BIDS Apps website lists a few dozen apps.[17]

What is particularly nice about this approach is that it does not necessarily require the developers of existing tools to do a lot of extra work themselves to support BIDS. In principle, anyone can write a BIDS App *wrapper* that mediates between the BIDS format and whatever format a tool natively expects to receive data in. So, for example, the BIDS Apps registry already contains BIDS Apps for packages or pipelines like SPM,[18] CPAC,[19] Freesurfer,[20] and the Human Connectome Project Pipelines.[21] Some of these apps are still fairly rudimentary and do not cover all of the functionality provided by the original tools, but others support much or most of the respective native tool's functionality. And of course, many BIDS Apps are not wrappers around other tools at all; they are entirely new tools designed from the very beginning to support only BIDS data sets. We have already mentioned fMRIPrep, which has very quickly become arguably the de facto

17. https://bids-apps.neuroimaging.io/apps/
18. https://www.fil.ion.ucl.ac.uk/spm/
19. https://fcp-indi.github.io/
20. https://surfer.nmr.mgh.harvard.edu/
21. https://www.humanconnectome.org/software/hcp-mr-pipelines

preprocessing pipeline in fMRI; another widely used BIDS App is MRIQC[22] [Esteban et al. 2017], a tool for automated quality control and quality assessment of structural and functional MRI scans, which we will see in action in chapter 12. Although the BIDS Apps ecosystem is still in its infancy, the latter two tools already represent something close to perfect applications for many researchers.

To demonstrate this statement, consider how easy it is to run fMRIPrep once your data are organized in the BIDS format. After installing the software and its dependencies, running the software is as simple as issuing this command in the terminal

```
fmriprep data/bids_root/ out/ participant -w work/
```

where `data/bids_root` points to a directory containing a BIDS-organized data set that includes fMRI data, `out` points to the directory into which the outputs (the BIDS derivatives) will be saved, and `work` is a directory that will store some of the intermediate products generated along the way. Looking at this, it might not be immediately apparent how important BIDS is for this to be so simple, but consider what software would need to do to find all of the fMRI data inside of a complex data set of raw MRI data, which might contain other data types, other files, and so forth. Consider also the complexity that arises from the fact that fMRI data can be collected using many different acquisition protocols, and the fact that fMRI processing sometimes uses other information (e.g., measurements of the field map, or anatomical T1-weighted or T2-weighted scans). The fact that the data comply with BIDS allows fMRIPrep to locate everything that it needs with the data set and to make use of all the information to perform the preprocessing to the best of its ability given the provided data.

UTILITY LIBRARIES

Lastly, the widespread adoption of BIDS has also spawned a large number of utility libraries designed to help developers (rather than end users) build their analysis pipelines and tools more efficiently. Suppose I am writing a script to automate my lab's typical fMRI analysis workflow. It is a safe bet that, at multiple points in my script, I will need to interact with the input data sets in fairly stereotyped and repetitive ways. For instance, I might need to search through the project directory for all files containing information about event timing, but only for a particular experimental task. Or, I might need to extract some metadata containing key parameters for each time series image I want to analyze (e.g., the repetition time, or TR). Such tasks are usually not very complicated, but they are tedious and can slow down development considerably. Worse, at a community level, they introduce massive inefficiency, because each person working on their analysis script ends up writing their own code to solve what are usually very similar problems.

A good utility library abstracts away a lot of this kind of low-level work and allows researchers and analysts to focus most of their attention on high-level objectives. By

---

22. https://mriqc.readthedocs.io/en/stable/

standardizing the representation of neuroimaging data, BIDS makes it much easier to write good libraries of this sort. Probably the best example so far is a package called PyBIDS,[23] which provides a convenient Python interface for basic querying and manipulation of BIDS data sets. To give you a sense of how much work PyBIDS can save you when you are writing neuroimaging analysis code in Python, let's take a look at some of the things the package can do.

We start by importing an object called BIDSLayout, which we will use to manage and query the layout of files on disk. We also import a function that knows how to locate some test data that were installed on our computer together with the PyBIDS software library.

```python
from bids import BIDSLayout
from bids.tests import get_test_data_path
```

One of the data sets that we have within the test data path is data set number 5 from OpenNeuro.[24] Note that the software has not actually installed a bunch of neuroimaging data onto our hard drive; that would be too large! Instead, the software installed a bunch of files that have the right names and are organized in the right way, but are mostly empty. This allows us to demonstrate the way that the software works, but note that you should not try reading the neuroimaging data from any of the files in that directory. We will work with a more manageable BIDS data set, including the files in it, in chapter 12.

For now, we initialize a BIDSLayout object, by pointing to the location of the data set in our disk. When we do that, the software scans through that part of our file system, validates that it is a properly organized BIDS data set, and finds all of the files that are arranged according to the specification. This allows the object to already infer some things about the data set. For example, the data set has sixteen subjects and forty-eight total runs (here a run is an individual fMRI scan). The person who organized this data set decided not to include a session folder for each subject. Presumably, because each subject participated in just one session in this experiment that information is not useful.

```python
layout = BIDSLayout(get_test_data_path() + "/ds005")
print(layout)
```

```
BIDS Layout: ...packages/bids/tests/data/ds005 | Subjects: 16 |
↪  Sessions: 0 |  Runs: 48
```

The layout object now has a method called get(), which we can use to gain access to various parts of the data set. For example, we can ask it to give us a list of the filenames of all of the anatomical ("T1w") scans that were collected for subjects sub-01 and sub-02

---

23. https://github.com/bids-standard/pybids
24. https://openneuro.org/datasets/ds000005

```
layout.get(subject=['01', '02'],  suffix="T1w", return_type='filename')
```

```
['/Users/arokem/miniconda3/envs/nds/lib/python3.8/site-packages/bids ⌋
 ↪    /tests/data/ds005/sub-01/anat/sub-01_T1w.nii.gz',
 '/Users/arokem/miniconda3/envs/nds/lib/python3.8/site-packages/bids ⌋
 ↪    /tests/data/ds005/sub-02/anat/sub-02_T1w.nii.gz']
```

Or, using a slightly different logic, we can ask for all of the functional ("bold") scans collected for subject sub-03

```
layout.get(subject='03', suffix="bold", return_type='filename')
```

```
['/Users/arokem/miniconda3/envs/nds/lib/python3.8/site-packages/bids ⌋
 ↪    /tests/data/ds005/sub-03/func/sub-03_task-mixedgamblestask_run-0 ⌋
 ↪    1_bold.nii.gz',
 '/Users/arokem/miniconda3/envs/nds/lib/python3.8/site-packages/bids ⌋
 ↪    /tests/data/ds005/sub-03/func/sub-03_task-mixedgamblestask_run- ⌋
 ↪    02_bold.nii.gz',
 '/Users/arokem/miniconda3/envs/nds/lib/python3.8/site-packages/bids ⌋
 ↪    /tests/data/ds005/sub-03/func/sub-03_task-mixedgamblestask_run- ⌋
 ↪    03_bold.nii.gz']
```

In these examples, we asked the BIDSLayout object to give us the 'filename' return type. This is because if we do not explicitly ask for a return type, we will get back a list of BIDSImageFile objects. For example, we can select the first one of these for sub-03's fMRI scans:

```
bids_files = layout.get(subject="03", suffix="bold")
bids_image = bids_files[0]
```

This object is quite useful. For example, it knows how to parse the file name into meaningful entities, using the get_entities() method, which returns a dictionary with entities such as subject and task that can be used to keep track of things in analysis scripts.

```
bids_image.get_entities()
```

```
{'datatype': 'func',
 'extension': '.nii.gz',
 'run': 1,
 'subject': '03',
 'suffix': 'bold',
 'task': 'mixedgamblestask'}
```

In most cases, you can also get direct access to the imaging data using the BIDSImageFile object. This object has a get_image method, which would usually return a NiBabel Nifti1Image object. As you will see in chapter 12 this object lets you extract metadata, or even read the data from a file into memory as a NumPy array. However, in this case, calling the get_image method would raise an error because, as we mentioned previously, the files do not contain any data. So, let's look at another kind of file that you can read directly in this case. In addition to the neuroimaging data, BIDS provides instructions on how to organize files that record the behavioral events that occurred during an experiment. These are stored as tab-separated value (TSV) files, and there is one for each run in the experiment. For example, for this data set, we can query for the events that happened during the third run for subject sub-03:

```
events = layout.get(subject='03', extension=".tsv", task="mixedgamblestask", run="03")
tsv_file = events[0]
print(tsv_file)
```

```
<BIDSDataFile filename='/Users/arokem/miniconda3/envs/nds/lib/python⌐
 ↪   3.8/site-packages/bids/tests/data/ds005/sub-03/func/sub-03_task-⌐
 ↪   mixedgamblestask_run-03_events.tsv'>
```

Instead of a BIDSImageFile, the variable tsv_file is now a BIDSDataFile object, and this kind of object has a get_df method, which returns a Pandas DataFrame object

```
bids_df = tsv_file.get_df()
bids_df.info()
```

```
<class 'pandas.core.frame.DataFrame'>
RangeIndex: 85 entries, 0 to 84
Data columns (total 12 columns):
 #   Column                   Non-Null Count   Dtype
---  ------                   --------------   -----
 0   onset                    85 non-null      float64
 1   duration                 85 non-null      int64
 2   trial_type               85 non-null      object
 3   distance from indiference  0 non-null     float64
 4   parametric gain          85 non-null      float64
 5   parametric loss          0 non-null       float64
 6   gain                     85 non-null      int64
 7   loss                     85 non-null      int64
 8   PTval                    85 non-null      float64
 9   respnum                  85 non-null      int64
 10  respcat                  85 non-null      int64
 11  RT                       85 non-null      float64
dtypes: float64(6), int64(5), object(1)
memory usage: 8.1+ KB
```

This kind of functionality is useful if you are planning to automate your analysis over large data sets that can include heterogeneous acquisitions between subjects and within subjects. At the very least, we hope that the examples have conveyed to you the power inherent in organizing your data according to a standard, as a starting point to use and maybe also develop, analysis pipelines that expect data in this format. We will see more examples of this in practice in the next chapter.

Exercise

BIDS has a set of example data sets available in a GitHub repository at https://github.com/bids-standard/bids-examples. Clone the repository and use PyBIDS to explore the data set called ds011. Using only PyBIDS code, can you figure out how many subjects participated in this study? What are the values of TR that were used in fMRI acquisitions in this data set?

11.3   Additional Resources

There are many places to learn more about the NiPy community. The documentation of each of the software libraries that were mentioned here includes fully worked-out examples of data analysis, in addition to links to tutorials, videos, and so forth. For Nipype, in particular, we recommend Michael Notter's Nipype tutorial[25] as a good starting point.

BIDS is not only a data standard but also a community of developers and users that support the use and further development of the standard. For example, since BIDS was first proposed, the community has added instructions to support the organization and sharing of new data modalities (e.g., intracranial EEG) and derivatives of processing. The strength of such a community is that, like open-source software, it draws on the strengths of many individuals and groups who are willing to spend time and effort evolving and improving it. One of the community-developed resources designed to help people who are new to the standard learn more is the BIDS Starter Kit.[26] It is a website that includes materials (videos, tutorials, explanations, and examples) to help you get started learning about and eventually using the BIDS standard.

In addition to these relatively static resources, users of both NiPy software and BIDS can interact with other members of the community through a dedicated questions and answers website called Neurostars.[27] On this website, anyone can ask questions about neuroscience software and get answers from experts who frequent the website. In particular, the developers of many of the projects that we mentioned in this chapter, and many of the people who work on the BIDS standard, often answer questions about these projects through this website.

---

25. https://miykael.github.io/nipype_tutorial/
26. https://bids-standard.github.io/bids-starter-kit/
27. https://neurostars.org/

12

Reading Neuroimaging Data with `NiBabel`

The first step in almost any neuroimaging data analysis is reading neuroimaging data from files. We will start dipping our toes into how to do this with the NiBabel software library. Many different file formats store neuroimaging data, but the NIfTI format (the acronym stands for Neuroimaging Informatics Technology Initiative, which was the group at the U.S. National Institutes of Health that originally designed the file format) has gained significant traction in recent years. For example, it is the format required by the Brain Imaging Data Structure (BIDS) standard for storing raw MRI data. Because this file format is so common, we will focus solely on NIfTI and not go into detail on the other file formats that are supported by NiBabel.

Getting acquainted with NiBabel and its operations will also allow us to work through some issues that arise when analyzing data from neuroimaging experiments. In the next chapter (chapter 13), we will dive even deeper into the challenge of aligning different measurements to each other. Let's start by downloading some neuroimaging files and seeing how NiBabel works. We will use the data downloading functionality implemented in our ndslib software library to get data from the OpenNeuro archive.[1] These two files are part of a study on the effects of learning on brain representations [Beukema et al. 2019]. From this data set, we will download just one run of a functional MRI (fMRI) experiment and a corresponding T1-weighted image acquired for the same subject in the same session.

```
from ndslib.data import download_bids_dataset
download_bids_dataset()
```

After running this code, we should have a valid BIDS data set inside a folder called `ds001233`. If you examine your file system, you will see that the files have been named and organized according to the BIDS standard. This will come in handy in a little bit.

---

1. https://openneuro.org/

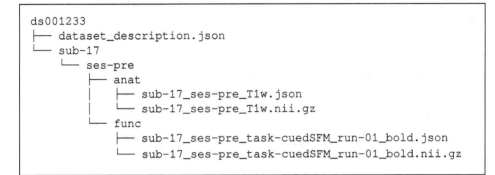

```
ds001233
├── dataset_description.json
└── sub-17
     └── ses-pre
          ├── anat
          │    ├── sub-17_ses-pre_T1w.json
          │    └── sub-17_ses-pre_T1w.nii.gz
          └── func
               ├── sub-17_ses-pre_task-cuedSFM_run-01_bold.json
               └── sub-17_ses-pre_task-cuedSFM_run-01_bold.nii.gz
```

In the meanwhile, let's start using NiBabel to access the data:

```
import nibabel as nib
img_bold = nib.load(" ⌐
  ↪  ds001233/sub-17/ses-pre/func/sub-17_ses-pre_task-cuedSFM_run-01_
  ↪  bold.nii.gz")
```

What kind of object is `img_bold`?

```
type(img_bold)
```

```
nibabel.nifti1.Nifti1Image
```

As you can see, it is a `Nifti1Image` object. This is a special NiBabel object for handling NIfTI files. One of the principles that guide the design of NiBabel is that operations
that take time, or that require a lot of the computer's memory are delayed and only executed when there is no choice. Instead, NiBabel starts by reading a relatively small section
of the file, called the file's header that contains only meta data: information about the data,
rather than the data itself. For example, the header contains information about the size of
the data array that is stored inside the file. So, when we called `nibabel.load` in the
previous example none of the actual data was read. Instead, NiBabel reads the header information about this data and could be prepared to also give you the data if and when you
want it. This is useful for cases where we need to know some property of the data, such as
the size of the data array, but we do not yet need to read any of the data itself. For example,

```
print(img_bold.shape)
```

```
(96, 96, 66, 241)
```

This tells us that the data in this file are a four-dimensional array with dimensions
of 96 × 96 × 66 voxels, collected in two hundred forty-one time points. There is other

information stored in the header, and we will look at some of this information a little bit later. First, we will use the object's get_fdata method to read the data from the file into memory. The 'f' in this method name indicates to us that NiBabel is going to take the data that is stored in the file and do its best to represent it as floating point numbers, regardless of the way the data are stored in the file.

```
data_bold = img_bold.get_fdata()
type(data_bold)
```

```
numpy.ndarray
```

Now it is a regular NumPy array. Everything that applies to the arrays that you already saw in previous chapters applies to this array. You can compute on it, visualize it, and so forth. What data type does it use?

```
print(data_bold.dtype)
```

```
float64
```

As promised, the data, regardless of how they were stored in the file, are given to us in the 64-bit floating point NumPy data type. There are other ways to read data from NiBabel, that allow you more fine-grained control of how much of the data you are reading, how and whether you store it in memory, and so on, but we will not go into that here. We recommend reading the NiBabel documentation to get acquainted with these details. One of the reasons that we do not think this is worth expounding on is that the get_fdata does the right thing for you in 99% of the cases that we have encountered ourselves. So, this provides a simple way to read the data and know what you are getting.

What shape does this array have?

```
print(data_bold.shape)
```

```
(96, 96, 66, 241)
```

It is exactly the shape you would expect from the information provided in the Nifti1Image that we saw previously. There are many things we can do with this data, but one of the first things that you might want to do with a new data set is to assess the quality of the data. This will help guide our analysis strategy and allow us to identify samples (time points, runs, subjects) that should be handled carefully (e.g., excluded from further analysis, or considered with a grain of salt).

12.1   Assessing MRI Data Quality

MRI data quality assessments are important because MRI data can vary quite a bit in terms of their quality and the conclusions from data analysis may be misguided if data quality is not taken into account. For example, the quality of measurements from young children and older adults could be affected by the tendency of participants from these populations to move more in the MRI scanner. There are many different ways to assess MRI data quality. We will start with a commonly used metric that is also relatively easy to understand, by computing the temporal signal-to-noise ratio (tSNR) of the measurement in each voxel. This is quantified as the mean signal in the voxel across time points, divided by the standard deviation of the signal across time points. To demonstrate the various concepts that we discussed previously, we will use three different ways to compute this value. The first uses NumPy directly on the data that we read from the file:

```
import numpy as np
tsnr_numpy = np.mean(data_bold, -1) / np.std(data_bold, -1)
```

The second way to compute relies on the Nipype software. We will not go into much depth about how to use the Nipype library, recommending instead that you refer to one of the resources provided subsequently. But briefly, Nipype interfaces are objects that provide direct Python control of a variety of software, including functionality that implemented in many non-Python libraries, such as AFNI and FSL. Here, we are going to use a Nipype interface for a relatively straightforward Python-based calculation of tSNR, very similar to the calculation we just did using NumPy. The way this is done here is by defining the interface object TSNR(), associating this object with an input file, the file that contains the fMRI data, and then calling the .run() method. This creates a file – tsnr.nii.gz – which contains the results of the computation.

```
from nipype.algorithms.confounds import TSNR

tsnr = TSNR()
tsnr.inputs.in_file =
 ↪  'ds001233/sub-17/ses-pre/func/sub-17_ses-pre_task-cuedSFM_run-⌐
 ↪  01_bold.nii.gz'
res = tsnr.run()
```

The output file (tsnr.nii.gz) can be immediately loaded with NiBabel:

```
tsnr_img = nib.load("tsnr.nii.gz")
tsnr_data = tsnr_img.get_fdata()
```

> **Exercise**
>
> Are the results from the NumPy calculation and the Nipype calculation identical? What happens if you exclude all the voxels for which the standard deviation is zero or very small (e.g., smaller than 0.001)?

A third option to calculate quality assessments for MRI data will use a BIDS App. The MRIQC app, mentioned in Section 11.2 is implemented as a Docker container. It takes a valid BIDS data set as input and produces a range of quality assessments as output. Let's see how this works in practice. In a terminal prompt, we first create a folder for the outputs of processing:

```
$ mkdir output
```

Then, we use Docker, much as we did in chapter 4, to point the MRIQC Docker container to the BIDS data set:

```
$ docker run -it --rm -v $(pwd)/ds001233:/data:ro -v
↪   $(pwd)/output:/out nipreps/mriqc:latest /data /out participant
↪   --participant_label 17
```

Because the data set is rather small—one fMRI run and one T1-weighted scan—it takes just a few minutes to run. When this is completed, we should have the following contents in the 'output' folder:

```
output
├── dataset_description.json
├── logs
├── sub-17
│   └── ses-pre
│       ├── anat
│       │   └── sub-17_ses-pre_T1w.json
│       └── func
│           └── sub-17_ses-pre_task-cuedSFM_run-01_bold.json
├── sub-17_ses-pre_T1w.html
└── sub-17_ses-pre_task-cuedSFM_run-01_bold.html
```

Notice that the outputs of this processing are also organized as a (derivatives) BIDS data set. There are no neuroimaging data files in here, but the files named `sub-17_ses-pre_T1w.json` and `sub-17_ses-pre_task-cuedSFM_run-01_bold.json` each contain a dictionary of quality assessment metrics, including for `sub-17_ses-pre_task-cuedSFM_run-01_bold.json`, a field with the median tSNR value that was calculated after applying motion correction, in addition to many other metrics. In addition, the HTML files saved at the top level of the outputs contain some visual diagnostics of image quality. This example again demonstrates the importance of the BIDS standard. Without it, the dockerized MRIQC BIDS App would not know how to find and understand the data that are stored in the input folder, and we would not necessarily have an immediate intuitive understanding of the outputs that were provided.

Exercise

Is the MRIQC median tSNR larger or smaller than the ones calculated with Nipype and NumPy? Why is that?

12.2 Additional Resources

There is extensive literature on approaches to quality assessment of MRI data. The MRIQC documentation[2] is a good place to start reading about the variety of metrics that can be used to automatically characterize image quality. In particular, to see some of the impacts of MRI quality assessments on the conclusions from MRI studies, you can look at some of the original research literature on this topic. Across different MRI modalities, researchers have shown that conclusions from MRI studies can be confounded particularly by the impact that motion has on MRI data quality. This has been identified in diffusion MRI [Yendiki et al. 2014], resting state functional connectivity [Power et al. 2012] and in the analysis of structural MRI [Reuter et al. 2015].

---

2. https://mriqc.readthedocs.io/

13

Using Nibabel to Align
Different Measurements

In the previous section, we saw how to load data from files using NiBabel. Next, we will go a little bit deeper into how we can use the metadata that are provided in these files. One of the main problems that we might encounter in analyzing MRI data is that we would like to combine information acquired using different kinds of measurements. For example, in the previous section, we read a file that contains functional MRI (fMRI) blood oxygenation level dependent (BOLD) data. We used the following code:

```
import nibabel as nib
img_bold = nib.load("ds001233/sub-17/ses-pre/func/sub-17_ses-pre_task-cuedS⌋
↪  FM_run-01_bold.nii.gz")
data_bold = img_bold.get_fdata()
```

In this case, the same person whose fMRI data we just loaded also underwent a T1-weighted scan in the same session. This image is used to find different parts of this person's anatomy. For example, it can be used to define where this person's gray matter is and where the gray matter turns into white matter, just underneath the surface of the cortex.

```
img_t1 = nib.load("ds001233/sub-17/ses-pre/anat/sub-17_ses-pre_T1w.nii.gz")
data_t1 = img_t1.get_fdata()
```

If we look at the metadata for this file, we will already get a sense that there might be a problem:

```
print(img_t1.shape)
```

```
(256, 256, 176)
```

The data is three-dimensional, and the three first dimensions correspond to the spatial dimensions of the fMRI data, but the shape of these dimensions does not correspond to the shape of the first three dimensions of the fMRI data. In this case, we happen to know that the fMRI was acquired with a spatial resolution of $2 \times 2 \times 2$ mm, and the T1-weighted data was acquired with a spatial resolution of $1 \times 1 \times 1$ mm. But even taking that into account, multiplying the size of each dimension of the T1 image by the ratio of the resolutions (a factor of two, in this case) does not make things equal. This is because the data was also acquired using a different field of view: the T1-weighted data covers 256×256 mm in each slice (and there are one hundred seventy-six slices, covering 176 mm), while the fMRI data covers only $96 \times 2 = 192$ mm by 192 mm in each slice (and there are sixty-six slices, covering only 132 mm). How would we then know how to match a particular location in the fMRI data to particular locations in the T1-weighted image? This is something that we need to do if we want to know if a particular voxel in the fMRI is inside of the brain's gray matter or in another anatomical segment of the brain.

In addition to the differences in acquisition parameters, we also need to take into account that the brain might be in a different location within the scanner in different measurements. This can be because the subject moved between the two scans, or because the two scans were done on separate days and the head of the subject was in a different position on these two days. The process of figuring out the way to bring two images of the same object taken when the object is in different locations is called *image registration*. We will need to take care of that too, and we will return to discuss this in Section 16. But, for our current discussion, we are going to assume that the subject's brain was in the same place when the two measurements were made. In the case of this particular data, this happens to be a good assumption, probably because these two measurements were done in close succession, and the subject did not move between the end of one scan and the beginning of the next one.

But even in this case, we still need to take care of *image alignment*, because of the differences in image acquisition parameters. Fortunately, the NIfTI file contains information that can help us align the two volumes to each other, despite the difference in resolution and field of view.

This information is stored in another piece of metadata in the header of the file, which is called the *affine matrix*. This is a 4×4 matrix that contains information that tells us how the acquisition volume was positioned relative to the MRI instrument, and relative to the subject's brain, which is inside the magnet. The affine matrix stored together with the data provides a way to unambiguously denote the location of the volume of data relative to the scanner.

13.1   Coordinate Frames

To understand how we will use this information, let's first look at a concrete example of the problem. For example, consider what happens when we take a single volume of the fMRI data, slice it on the last dimension so that we get the middle of the volume on that dimension, and compare it to the middle slice of the last dimension of the T1-weighted data.

```
data_bold = img_bold.get_fdata()

data_bold_t0 = data_bold[:, :, :, 0]

import matplotlib.pyplot as plt
fig, ax = plt.subplots(1, 2)
ax[0].matshow(data_bold_t0[:, :, data_bold_t0.shape[-1]//2])
im = ax[1].matshow(data_t1[:, :, data_t1.shape[-1]//2])
```

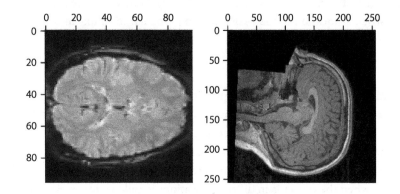

These two views of the subject's brain are very different from each other. The data are not even oriented the same way in terms of the order of the anatomical axes! While the middle slice on the last spatial dimension for the fMRI data gives us an axial slice from the data, for the T1-weighted data, we get a sagittal slice.

To understand why that is the case, we need to understand that the coordinate frame in which the data is collected can be quite different in different acquisitions. What do we mean by *coordinate frame*? Technically, a coordinate frame is defined by an origin and axes emanating from that origin, as well as the directions and units used to step along the axes. For example, when we make measurements in the MRI instruments, we define a coordinate frame of the scanner as one that has the iso center of the scanner as its origin. This is the point in the scanner in which we usually place the subject's brain when we make measurements. We then use the way in which the subject is usually placed into the MRI bore to explain the convention used to define the axes. In neuroimaging, this is usually done by placing the subject face up on a bed and then moving the bed into the magnet with the head going into the MRI bore first (this is technically known as the "supine, head first" position). Using the subject's position on the bed, we can define the axes based on the location of the brain within the scanner. For example, we can define the first (x) axis to be one that goes from left to right through the subject's brain as they are lying inside the MRI bore, and the second (y) axis is defined to go from the floor of the MRI room, behind the back of the subject's brain through the brain and up through the front of their head toward the ceiling of the MRI room. In terms of the subject's head, as they are placed in the supine, head first position, this is the *anterior-posterior* axis. The last (z) axis goes from their feet lying outside the magnet, through the length of their body, and through the top of their head (this is *inferior-superior* relative to their anatomy). Based on these axes, we say that this coordinate frame has the right-anterior-superior (RAS) orientation. This means that the coordinates increase toward the right side of the subject, toward the anterior part of their brain, and toward the superior part of their head.

In the spatial coordinate frame of the MRI magnet, we often use millimeters to describe how we step along the axes. So, specific coordinates in millimeters point to particular points within the scanner. For example, the coordinate [1, 0, 0] is 1 mm to the right of the iso center (where right is defined as the direction to the subject's right as they are lying down). Following this same logic, negative coordinates can be used to indicate points that are to the left, posterior, and inferior of the scanner iso center (again, relative to the subject's brain). So, the coordinate [-1, 0, 0] indicates a point that is 1 mm to the left of the iso center. The RAS coordinate system is a common way to define the scanner coordinate frame, but you can also have LPI (which goes right-to-left, anterior-to-posterior, and superior-to-inferior instead) or any combination of directions along these principal directions (RAI, LPS, etc.).

Exercise

What is the location of [1, 1, 1] RAS in the LAS coordinate frame?

Given that we know where particular points are in space relative to the magnet iso center, how can we tell where these particular points are in the NumPy array that represents our data? For example, how would we translate the coordinate [1, 0, 0] in millimeters into an index [i, j, k], which we can use to index into the array to find out what the value of the MRI data is in that location?

The volume of data is acquired with a certain orientation, a certain field of view, and a certain voxel resolution. This means that the data also has a spatial coordinate frame. The [0, 0, 0] coordinate indicates the voxel that is the array coordinate [0, 0, 0] (i.e., i=0, j=0, and k=0) and the axes are the axes of the array. For example, the fMRI data we have been looking at here is organized such that [0, 0, 0] is somewhere inferior and posterior to their head. We can tell that by plotting several slices of the brain starting at slice 20, going to the middle slice, and then up to slice 46 (twenty slices from the end of the array in that direction).

```
import matplotlib.pyplot as plt
fig, ax = plt.subplots(1, 3)
ax[0].matshow(data_bold_t0[:, :, 20])
ax[1].matshow(data_bold_t0[:, :, data_bold_t0.shape[-1]//2])
ax[2].matshow(data_bold_t0[:, :, 46])
fig.set_tight_layout("tight")
```

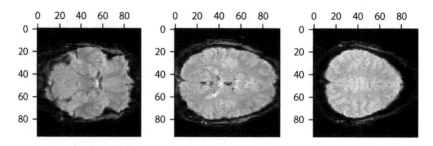

Based on our knowledge of anatomy, we see that the first slice is lower in the head than the middle slice, which is lower in the head than the third slice. This means that as we increase the coordinate on the last dimension we are going up in the brain. So, we know at least that the last coordinate is an "S." Based on the labeling of the axis, we can also see that the second dimension increases from the back of the brain to the front of the brain, meaning that the second coordinate is an "A." So far, so good. What about the first dimension? We might be tempted to think that since it looks like it is increasing from left to right, which must be an "R." But this is where we need to pause for a bit. This is because there is nothing to tell us whether the brain is right-left flipped. This is something that we need to be particularly careful with (right-left flips are a major source of error in neuroimaging data analysis).

Exercise

Based on the anatomy, what can you infer about the coordinate frame in which the T1-weighted data was acquired?

We see that we can guess our way toward the coordinate frame in which the data was acquired, but as we already mentioned previously, there is one bit of data that will tell you unambiguously how to correspond between the data and the scanner coordinate frame. To use this bit of data to calculate the transformation between different coordinate frames, we are going to use a mathematical operation called matrix multiplication. If you have never seen this operation before, you can learn about it in the section that follows. Otherwise, you can skip on to Section 13.3.

13.1.1 Matrix Multiplication Matrices, Vectors and Their Multiplication

A matrix is a collection of numbers that are organized in a table with rows and columns. Matrices look a lot like two-dimensional NumPy arrays, and NumPy arrays can be used to represent matrices.

In the case where a matrix has only one column or only one row, we call it a vector. If it has only one column and only one row; i.e., it is just a single number, we refer to it as a scalar.

Let's look at an example. The following is a 2×3 matrix:

$$A = \begin{bmatrix} a & b & c \\ d & e & f \end{bmatrix}$$

The following is a three-vector, with one row (we will use a convention where matrices are denoted by capital letters and vectors are denoted by lowercase letters):

$$v = \begin{bmatrix} a & b & c \end{bmatrix}$$

We can also write the same values into a vector with three rows and one column.

$$v = \begin{bmatrix} a \\ b \\ c \end{bmatrix}$$

In the most general terms, the matrix multiplication between the $m \times n$ matrix A and the $n \times p$ matrix B result in an $m \times p$ matrix. The entry on row i and column j in the resulting matrix, C is defined as

$$C_{i,j} = \sum_{k=1}^{n} A_{i,k} B_{k,j}$$

If this looks a bit mysterious, let's consider an example and work out the computations of individual components. Let's consider

$$A = \begin{bmatrix} 1 & 2 & 3 \\ 4 & 5 & 6 \end{bmatrix}$$

$$B = \begin{bmatrix} 7 & 8 \\ 9 & 10 \\ 11 & 12 \end{bmatrix}$$

The first matrix A is a 2×3 matrix, and the second matrix, B is a 3×2 matrix. Based on that, we already know that the resulting multiplication, which we write as $C = AB$ will be a 2×2 matrix.

Let's calculate the entries in this matrix one by one. Confusingly enough, in contrast to Python sequences, when we do math with matrices, we index them using a one-based indexing system. This means that the first item in the first row of this matrix is $C_{1,1}$. The value of this item is

$$C_{1,1} = \sum_{k=1}^{3} A_{1,k} B_{k,1}$$

$$= A_{1,1} B_{1,1} + A_{1,2} B_{2,1} + A_{1,3} B_{3,1}$$

$$= 1 \cdot 7 + 2 \cdot 9 + 3 \cdot 11 = 58$$

Similarly, the value of $C_{1,2}$ is

$$C_{1,2} = \sum_{k=1}^{3} A_{1,k} B_{k,2}$$

$$= A_{1,1}B_{1,2} + A_{1,2}B_{2,2} + A_{1,3}B_{3,2}$$

$$= 1 \cdot 8 + 2 \cdot 10 + 3 \cdot 12 = 64$$

And so on. At the end of it all, the matrix C will be

$$C = \begin{bmatrix} 58 & 64 \\ 139 & 154 \end{bmatrix}$$

Exercises

1. Convince yourself that the answer we gave previously is correct by computing the components of the second row following the formula.
2. Write a function that takes two matrices as input and computes their multiplication.

13.2   Multiplying Matrices in Python

Fortunately, NumPy already implements a (very fast and efficient!) matrix multiplication for matrices and vectors that are represented with NumPy arrays. The standard multiplication sign * is already being used in NumPy for element-by-element multiplication (see chapter 8), so another symbol was adopted for matrix multiplication: @. As an example, let's look at the matrices A, B, and C we used previously[1]

```python
import numpy as np
A = np.array([[1, 2, 3], [4, 5, 6]])

B = np.array([[7, 8], [9, 10], [11, 12]])

C = A @ B

print(C)
```

```
[[ 58  64]
 [139 154]]
```

13.3   Using the Affine

The way that the affine information is used requires a little bit of knowledge of matrix multiplication. If you are not at all familiar with this idea, a brief introduction is given in

---

1. There is also a function that implements this multiplication: `np.dot(A, B)` is equivalent to `A @ B`

the previous section that includes pretty much everything you need to know about this topic to follow along from this point forward.

13.3.1   Coordinate Transformations

One particularly interesting case of matrix multiplication is the multiplication of vectors of length n with an $n \times n$ matrix. This is because the input and the output in this case have the same number of entries. In other words, the dimensionality of the input and output are identical. One way to interpret this kind of multiplication is to think of it as taking one coordinate in a space defined by n coordinates and moving it to another coordinate in that n-dimensional space. This idea is used as the basis for transformations between the different coordinate spaces that describe where our MRI data is.

Let's think about some simple transformations of coordinates. One kind of transformation is a scaling transformation. This takes a coordinate in the n-dimensional space (e.g., in the two-dimensional space [x, y]) and sends it to a scaled version of that coordinate (e.g., to [a*x, b*y]). This transformation is implemented through a diagonal matrix. Here is the two-dimensional case:

$$\begin{bmatrix} a & 0 \\ 0 & b \end{bmatrix}$$

If you are new to matrix multiplications, you might want to convince yourself that this matrix does indeed implement the scaling, by working through the math of this example element by element. Importantly, this idea is not limited to two dimensions, or any particular dimensionality for that matter. For example, in the three-dimensional case, this would be

$$\begin{bmatrix} a & 0 & 0 \\ 0 & b & 0 \\ 0 & 0 & c \end{bmatrix}$$

and it would scale [x, y, z] to [a*x, b*y, c*z].

Another interesting transformation is what we call a rotation transformation. This takes a particular coordinate and rotates it around an axis.

$$A_\theta = \begin{bmatrix} \cos(\theta) & -\sin(\theta) \\ \sin(\theta) & \cos(\theta) \end{bmatrix}$$

What does it mean to rotate a coordinate around the axis? Imagine that the coordinate [x, y] is connected to the origin with a straight line. When we multiply that coordinate by our rotation matrix A_θ, we are rotating that line around the axis by that angle. It is easier to understand this idea through an example with code, and with a visual. The code that follows defines an angle theta (in radians) and the two-dimensional rotation matrix A_theta, and then multiplies a two-dimensional coordinate xy (plotted as a circle marker), which produces a new coordinate `xy_theta` (plotted with a square marker). We connect each coordinate to the origin (the coordinate `[0, 0]`) through a line: the original coordinate with a solid line and the new coordinate with a dashed line. The angle between these two lines is θ.

```
theta = np.pi/4
A_theta = np.array([[np.cos(theta), -np.sin(theta)], [np.sin(theta), np.cos(theta)]])
xy = np.array([2, 3])
xy_theta = A_theta @ xy

fig, ax = plt.subplots()
ax.scatter(xy[0], xy[1])
ax.plot([0, xy[0]], [0, xy[1]])
ax.scatter(xy_theta[0], xy_theta[1], marker='s')
ax.plot([0, xy_theta[0]], [0, xy_theta[1]], '--')
p = ax.axis("equal")
```

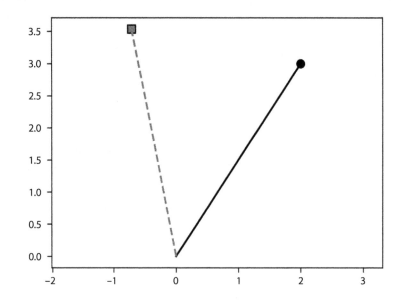

This idea can be expanded to three dimensions, but in this case, we define a separate rotation matrix for each of the axes. The rotation by angle θ around the x-axis is defined as

$$A_{\theta,x} = \begin{bmatrix} 1 & 0 & 0 \\ 0 & \cos(\theta) & -\sin(\theta) \\ 0 & \sin(\theta) & \cos(\theta) \end{bmatrix}$$

The rotation by angle θ around the y-axis is defined as

$$A_{\theta,y} = \begin{bmatrix} \cos(\theta) & 0 & -\sin(\theta) \\ 0 & 1 & 0 \\ \sin(\theta) & 0 & \cos(\theta) \end{bmatrix}$$

This leaves the rotation by angle θ around the z-axis, defined as

$$A_{\theta,z} = \begin{bmatrix} \cos(\theta) & -\sin(\theta) & 0 \\ \sin(\theta) & \cos(\theta) & 0 \\ 0 & 0 & 1 \end{bmatrix}$$

13.3.2 Composing Transformations

The transformations defined by matrices can be *composed* into new transformations. This is a technical term that means that we can do one transformation followed by another transformation, and that would be the same as multiplying the two transformation matrices by each other and then using the product. In other words, matrix multiplication is associative:

$$(AB)v = A(Bv)$$

For example, we can combine scaling and rotation.

```
theta = np.pi/4
A_theta = np.array([[np.cos(theta), -np.sin(theta)], [np.sin(theta), np.cos(theta)]])
B = np.array([[2, 0], [0, 2]])
xy = np.array([2, 3])
xy_transformed = A_theta @ B @ xy

fig, ax = plt.subplots()
ax.scatter(xy[0], xy[1])
ax.plot([0, xy[0]], [0, xy[1]])
ax.scatter(xy_transformed[0], xy_transformed[1], marker='s')
ax.plot([0, xy_transformed[0]], [0, xy_transformed[1]], '--')
p = ax.axis("equal")
```

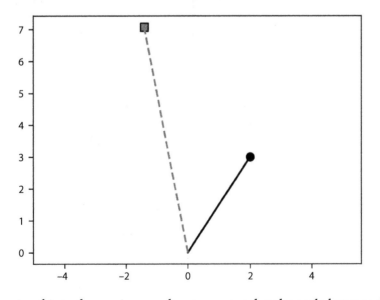

Comparing this to the previous result, you can see that the angle between the solid line and the dashed line is the same as it was in our previous example, but the length of the dashed line is now twice as long as the length of the solid line. This is because we applied both scaling and rotation. Notice also that following the associativity rule that we describe previously the line `xy_transformed` = `A_theta` @ `B` @ `xy` could be written `xy_transformed` = `(A_theta` @ `B)` @ `xy` or `xy_transformed` = `A_theta` @ `(B` @ `xy)` and the result would be the same.

Another thing to consider is that every coordinate in the two-dimensional space that we would multiply with this matrix would undergo the same transformation, relative to the origin: it would be rotated counterclockwise by an angle theta and moved to be twice as far from the origin as it previously was. In other words, the matrix performs a *coordinate transformation*.

Now, using this idea, we can start thinking about how we might describe data represented in one coordinate frame—the scanner space measured in millimeters—to another coordinate frame, i.e., the representation of the data in the array of acquired MRI data. This is done using a matrix that describes how each `[x,y,z]` position in the scanner space is transformed in an `[i,j,k]` coordinate in the array space.

Let's think first about the scaling: if each voxel is $2 \times 2 \times 2$ mm in size, that means that for each millimeter that we move in space, we are moving by half a voxel. A matrix that would describe this scaling operation would look like this:

$$A_{scale} = \begin{bmatrix} 0.5 & 0 & 0 \\ 0 & 0.5 & 0 \\ 0 & 0 & 0.5 \end{bmatrix}$$

The next thing to think about is the rotation of the coordinates: this is because when we acquire MRI data, we can set the orientation of the acquisition volume to be different from the orientation of the scanner coordinate frame. It can be tilted relative to the scanner axes in any of the three dimensions, which means that it could have rotations around each of the axes. So, we would combine our scaling matrix with a series of rotations.

$$A_{total} = A_{scale} A_{\theta,x} A_{\theta,y} A_{\theta,z}$$

But even after we multiply each of the coordinates in the brain in millimeters by this matrix, there is still one more operation that we need to do to move to the space of the MRI acquisition. As we mentioned previously, the origin of the scanner space is in the iso center of the magnet. But the MRI acquisition usually has its origin somewhere outside of the subject's head, so that it covers the entire brain. So, we need to move the origin of the space from the iso center to the corner of the scanned volume. Mathematically, this translates to shifting each coordinate `[x, y, z]` in the scanner space by $[\Delta x \Delta y \Delta z]$, where each one of these components describes where the corner of the MRI volume is relative to the position of iso center.

Unfortunately, this is not an operation that we can do by multiplying the coordinates with a 3×3 matrix of the sort that we saw previously. Instead, we will need to use a mathematical trick: instead of describing each location in the space via a three-dimensional coordinate `[x, y, z]`, we will add a dimension to the coordinates, so that they are written as `[x, y, z, 1]`. We then also make our matrix a 4×4 matrix

$$\begin{bmatrix} & & & \Delta x \\ & A_{total} & & \Delta y \\ & & & \Delta y \\ 0 & 0 & 0 & 1 \end{bmatrix}$$

where A_{total} is the matrix composed of our rotations and scalings. This might look like a weird thing to do, but an easy way to demonstrate that this shifts the origin of the space is to apply a matrix like this to the origin itself, which we have augmented with a 1 as the last element: `[0, 0, 0, 1]`. To make the example more straightforward, we will also set A_{total} to be a matrix with ones in the elements on the diagonal and all other elements set to 0. This matrix is called an *identity matrix* because when you multiply an n-dimensional vector with a matrix that is constructed in this manner, the result is identical to the original vector (use the formula in section 13.1.1, or create this matrix with code to convince yourself of that).

```
A_total = np.array([[1, 0, 0], [0, 1, 0], [0, 0, 1]])
delta_x = 128
delta_y = 127
delta_z = 85.5
A_affine = np.zeros((4, 4))
A_affine[:3, :3] = A_total
A_affine[:, 3] = np.array([delta_x, delta_y, delta_z, 1])

origin = np.array([0, 0, 0, 1])

result = A_affine @ origin
print(result)
```

```
[128.   127.    85.5   1. ]
```

As you may have noticed, in the code, we called this *augmented transformation matrix* an *affine* matrix. We will revisit affine transformations later in the book (when we use these transformations to process images in chapter 16). For now, we simply recognize that this is the name used to refer to the 4×4 matrix that is recorded in the header of neuroimaging data and tells you how to go from coordinates in the scanner space to coordinates in the image space. In NiBabel, these transformations are accessed via:

```
affine_t1 = img_t1.affine
affine_bold = img_bold.affine
```

Where do these affines send the origin?

```
print(affine_t1 @ np.array([0, 0, 0, 1]))
```

```
[ -85.5   128.   -127.      1. ]
```

This tells us that the element indexed as `data_t1[0, 0, 0]` lies 85.5 mm to the left of the iso center, 128 mm anterior to the iso center (toward the ceiling of the room) and 127 mm inferior to the iso center (toward the side of the room where the subject's legs are sticking out of the MRI).

Exercise

Where in the scanner is the element `data_bold[0, 0, 0]`? Where is `data_bold[-1, -1, -1]`? (Hint: `affine_bold @ np.array([-1, -1, -1, 1]` is not going to give you the correct answer).

13.3.3   The Inverse of the Affine

The *inverse* of a matrix is a matrix that does the opposite of what the original matrix did. The inverse of a matrix is denoted by adding a "−1" in the exponent of the matrix. For example, for the scaling matrix

$$B = \begin{bmatrix} 2 & 0 \\ 0 & 2 \end{bmatrix}$$

the inverse of B is called B^{-1}. Since this matrix sends a coordinate to be twice as far from the origin, its inverse would send a coordinate to be half as far from the origin. Based on what we already know, this would be

$$B^{-1} = \begin{bmatrix} 0.5 & 0 \\ 0 & 0.5 \end{bmatrix}$$

Similarly, for a rotation matrix

$$A_\theta = \begin{bmatrix} \cos(\theta) & -\sin(\theta) \\ \sin(\theta) & \cos(\theta) \end{bmatrix}$$

The inverse is the matrix that rotates the coordinate around the same axis by the same angle, but in the opposite direction

$$A_{\theta-1} = \begin{bmatrix} \cos(-\theta) & -\sin(-\theta) \\ \sin(-\theta) & \cos(-\theta) \end{bmatrix}$$

Exercise

The inverse of a rotation matrix is also its *transpose*, the matrix that has the rows and columns exchanged. Use the math you have seen so far (Hint: and also a bit of trigonometry) to demonstrate that this is the case.

13.3.4 Computing the Inverse of an Affine in Python

It will come as no surprise that there is a way to compute the inverse of a matrix in Python. That is the NumPy function: `np.linalg.inv`. As you can see from the name it comes from the `linalg` submodule (`linalg` stands for linear algebra, which is the field of mathematics that deals with matrices and operations on matrices). For example, the inverse of the scaling matrix B is computed as follows:

```
B_inv = np.linalg.inv(B)
print(B_inv)
```

```
[[0.5 0. ]
 [0.  0.5]]
```

Why do we need the inverse of a neuroimaging data file's affine? That is because we already know how to go from one data file's coordinates to the scanner space, but if we want to go from one data file's coordinates to another data file's coordinates, we will need to go from one data file to the scanner and then from the scanner coordinates to the other data file's coordinates. The latter part requires the inverse of the second affine.

For example, consider what we need to do if we want to know where the central voxel in the fMRI acquisition falls in the space of the T1-weighted acquisition. First, we calculate the location of this voxel in the `[i, j, k]` coordinates of the fMRI image space:

```
central_bold_voxel = np.array([img_bold.shape[0]//2,
                               img_bold.shape[1]//2,
                               img_bold.shape[2]//2, 1])
```

Next, we move this into the scanner space, using the affine transform of the fMRI image:

```
bold_center_scanner = affine_bold @ central_bold_voxel
print(bold_center_scanner)
```

```
[  2.          -19.61000896  16.75154221   1.         ]
```

Again, to remind you, these coordinates are now in millimeters, in scanner space. This means that this voxel is located 2 mm to the right, almost 20 mm posterior, and almost 17 mm superior to the MRI iso center.

Next, we use the *inverse* of the T1-weighted affine to move this coordinate into the space of the T1-weighted image space.

```
bold_center_t1 = np.linalg.inv(affine_t1) @ bold_center_scanner
print(bold_center_t1)
```

```
[147.61000896 143.75154221  87.5         1.        ]
```

This tells us that the time series in the central voxel of the fMRI volume comes approximately from the same location as the T1-weighted data in `data_t1[147 143 87]`. Because the data are sampled a little bit differently and the resolution differs, the boundaries between voxels are a little bit different in each of the recorded modalities. If you want to put them in the same space, you will also need to compensate for that. We will not show this in detail, but you should know that the process of alignment uses forms of data interpolation or smoothing to make up for these discrepancies, and these usually have a small effect on the values in the data. For example, resampling the data from the fMRI space to the T1 space and then back would not necessarily give you the same values you started with.

Taking all these facts together, we are now ready to align our data with each other. Fortunately, NiBabel implements a function that allows us to take one NIfTI image and resample it into the space of another image. This function takes a `NiftiImage1` object input and the shape and affine matrix of the space into which you would like to resample the data. So, we can go ahead and resample the T1-weighted image to the space and resolution of the fMRI data:

```
from nibabel.processing import resample_from_to
img_t1_resampled = resample_from_to(img_t1, (img_bold.shape[:3], img_bold.affine))
```

The image that results from this computation has the shape of the fMRI data, as well as the affine of the fMRI data:

```
print(img_t1_resampled.shape)
print(img_t1_resampled.affine == img_bold.affine)
```

```
(96, 96, 66)
[[ True   True   True   True]
 [ True   True   True   True]
 [ True   True   True   True]
 [ True   True   True   True]]
```

And, importantly, if we extract the data from this image, we can show that the two data modalities are now well aligned:

```
data_t1_resampled = img_t1_resampled.get_fdata()

fig, ax = plt.subplots(1, 2)
ax[0].matshow(data_bold_t0[:, :, data_bold_t0.shape[-1]//2])
im = ax[1].matshow(data_t1_resampled[:, :, data_t1_resampled.shape[-1]//2])
```

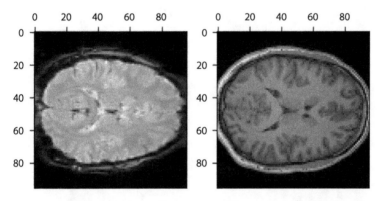

Finally, if you would like to write out this result into another NIfTI file to be used later in some other computation, you can do so by using NiBabel's save function:

```
nib.save(nib.Nifti1Image(data_t1_resampled, img_t1_resampled.affine),
         't1_resampled.nii.gz')
```

Exercises

1. Move the first volume of the fMRI data into the T1-weighted space and show that they are well aligned in that space. Save the resulting data into a new NIfTI file.
2. In which direction would you expect to lose more information to interpolation/smoothing? Going from the fMRI data to the T1-weighted data, or the other way around?

13.4   Additional Resources

Another explanation of coordinate transformations is provided in The NiBabel documentation.[2]

---

2. https://nipy.org/nibabel/coordinate_systems.html

PART V

Image Processing

14

Image Processing

Images are used as data in many different contexts. Numerous research fields use imaging instruments to study a variety of natural phenomena: these range from telescopes that image faraway galaxies, to powerful microscopes that can image individual molecular components of a living cell. But images are also used in commercial applications of data science, for example, in sorting through the photographs that we upload into social media or photo sharing services. Computer vision, another name for some of the advanced applications of image processing, has been used for a long time in things like quality control of printed circuit boards but is also frequently used in many new technologies, such as in providing driverless cars with the input they need to decide where to go.

Neuroscience research has been fascinated with images for almost as long as it has existed as a science. Santiago Ramón y Cajal, a Spanish neuroanatomist who lived around the turn of the twentieth century, and who was arguably one of the first modern neuroscientists, conducted a large part of his research by peering at brain tissue through a microscope and making often strikingly beautiful pictures of what he saw (you can see some of them on the Marginalian website[1]). One could argue that we have not made much progress, judging from the fact that a lot of contemporary research in neuroscience aims to do something very similar. More optimistically, we could say that the rapid development of imaging instruments and experimental techniques that has happened even just in the last few years has vaulted the field forward to create imaging data that show the activity of hundreds and even thousands of neurons simultaneously while animals are performing a specific behavior. It is still images, but they contain much more information than what Cajal had under his microscope. And we have fast computers and modern algorithms to analyze the data with, so maybe there is some progress.

Neuroimaging is of course all about images, and the fundamental type of data that most neuroimaging researchers use are images of brains. In this section, we will give a broad introduction to image processing that will survey a wide range of different kinds of operations that you can do with images. We will focus on the open-source Scikit Image[2] library and the operations that are implemented within this library. One of the advantages of using

---

1. https://www.themarginalian.org/2017/02/23/beautiful-brain-santiago-ramon-y-cajal/
2. https://scikit-image.org/

Scikit Image is that it is built as a domain-agnostic tool. Thus, instead of focusing on one particular application domain, the library exemplifies the interdisciplinary and quantitative nature of image processing by collecting together algorithms that have been developed in many different contexts.

Overall, the purpose of this chapter is to develop an intuition for the kind of things that you might think about when you are doing image processing. In the next chapters, we will zoom in and dissect in more detail two kinds of operations: image segmentation and image registration. These are particularly common in the analysis of neuroimaging data and a detailed understanding of these operations should serve you well, even if you ultimately use specialized off-the-shelf software tools for these operations, rather than implementing them yourself.

14.1   Images Are Arrays

As you saw earlier in this book, data from neuroimaging experiments can be represented as NumPy arrays. That is also true of images in general. So, what is the difference between an image and another kind of array? One thing that distinguishes images from other kinds of data is that spatial relationships are crucially important for the analysis of images. That is because neighboring parts of the image usually represent objects that are near each other in the physical world as well. As we will see subsequently, many kinds of image-processing algorithms use these neighborhood relationships. Conversely, in some cases, these neighborhood relationships induce constraints on the operations one can do with the images because we would like to keep these relationships constant as we put the image through various transformations. This too will come up when we discuss image processing methods that change the content of images.

14.2   Images Can Have Two Dimensions or More

We typically think of images as having two dimensions, because when we look at images we are usually viewing them on a two-dimensional surface. i.e., on the screen of a computer or a page. But some kinds of images have more than one dimension. For instance, when we are analyzing brain images, we are interested in the full three-dimensional image of the brain. This means that brain images are usually (at least) three-dimensional. Again, under our definition of an image, this is fine. So long as the spatial relationships are meaningful and important in the analysis, it is acceptable if they extend over more than two dimensions.

14.3   Images Can Have Other Special Dimensions

In some cases, we would like for there to be ways to represent several different properties of one location in the image. One common example of this is that images containing color need to specify three different values in each pixel for the color to be defined for that pixel.

Typically, these three values will represent the intensities of red (R), green (G), and blue (B) in the location. When we organize our data with another dimension that represents so-called color channels, the order of the channels is not necessarily meaningful; you can represent color images as RGB, as BGR, or even as GRB. So long as you use a consistent order, the meaning of the image does not change. Another interesting case of a special dimension is the case in which images of the same object were taken as a series in time. In this case, one dimension of the image can be time. In contrast to channels, time is similar to the spatial organization of the image in that the order of the images along this axis does matter. The fact that they were acquired in time makes the different samples of the brain amenable to a particular set of analysis methods called time series analysis, which is beyond the scope of our book.

To see this more clearly, let's get started with Scikit Image. The library includes a submodule that provides access to some image data sets that are used for testing and demonstrating the library's functionality. For example, one of the data sets is called `astronaut`. Let's load it and explore its properties.

```
import skimage
img = skimage.data.astronaut()
print(img.shape)
print(img.dtype)
```

```
(512, 512, 3)
uint8
```

The astronaut image is two-dimensional, but the array has three dimensions: this is because it has 512×512 pixels and each pixel stores three numbers corresponding to the R, G, and B color channels. These numbers are stored as 8-bit unsigned integers. This means that in each channel within each pixel the intensity of luminance at that location and that color is stored as a number between zero (no luminance in that color at all) to two hundred fifty-five (full luminance at that color). Let's use Matplotlib to visualize the contents of this image. In addition to the `matshow` function that you saw in chapter 10, Matplotlib implements an `imshow` function. One of the main differences between these is that `imshow` knows that images sometimes have color channels, and knows how to properly handle images that have three dimensions, such as this one. We also highlight two specific pixels in the image, by plotting white circular markers on top of them (for a color version of this image, please visit https://press.princeton.edu/book/data-science-for-neuroimaging).

```
import matplotlib.pyplot as plt
fig, ax = plt.subplots()
ax.imshow(img)
ax.plot(70, 200, marker='o', markersize=5, color="white")
p = ax.plot(200, 400, marker='o', markersize=5, color="white")
```

It turns out that this is a picture of Eileen Collins, who was the first woman to pilot a space shuttle and the first female Commander of a space shuttle. Apart from serving as a reminder of her inspiring story, and of the marvelous achievements of science, the image is a good example for exploring algorithms that analyze specific aspects of images. Many image processing algorithms are designed to detect faces in images (because faces are a particularly interesting object for many applications of image processing), and this image is useful because it contains a face.

Looking at specific pixels in the image, we see what it means for there to be a representation of color in the last dimension. Let's look at the two pixels that were highlighted in the image. The pixel at img[200, 70] is immediately behind Commander Collins' right shoulder inside one of the red stripes in the U.S. flag. The values of that pixel are relatively high for the first (red) channel and lower for the other two. That is the reason this pixel appears red. The pixel img[400, 200] is inside a patch on her space suit and is relatively highly weighted on the last (blue) channel. That is why it appears blue.

```
print(img[200, 70])
print(img[400, 200])
```

```
[122  14  29]
[ 79  36 126]
```

14.4   Operations with Images

Because of the information that is stored in the structure of image data, they are particularly amenable to certain kinds of operations. We will briefly look at a few of these here, and then look at a couple of these operations in depth in the next chapters.

14.4.1 Filtering

Filters are operations that take an image as an input and produce another image as the output. As their name suggests, the result of this kind of operation often retains some properties of the original image, while throwing away other properties of the image. For example, we can filter out the color, turning the image into a grayscale image:

```
from skimage.color import rgb2gray
gray_img = rgb2gray(img)
print(gray_img.shape)
```

```
(512, 512)
```

As you can see, this operation retains the pixel dimensions, but reduces the number of values in each pixel to one, instead of the three channel values that used to be there. In each pixel, the values in the different channels are combined using an equation that accounts for the sensitivity of human vision to changes in luminance in the different colors. The imshow function understands that because the image does not have a third dimension it should be displayed as a pseudocolor image. Here, it is displayed using a grayscale colormap. High values of gray are mapped to bright numbers and low values are mapped to dark colors.

```
import matplotlib.pyplot as plt
fig, ax = plt.subplots()
im = ax.imshow(gray_img)
```

Exercise

What are the values of the pixels that were previously highlighted in the grayscale image? Are these values high or low? Why do you think they get converted in this way? Take a look at the documentation for `rgb2gray` to get a hint.

14.4.2   Convolution Filters

There are many other kinds of filters that you can apply to an image. A large family of filters involves an operation called *convolution*. We will explain this operation here in some detail. This is both because understanding how a convolution works will help you understand some of the basics of image processing much better, but also because we will come back to this operation much later in the book when we discuss a family of machine learning algorithms called convolutional neural networks, which are particularly powerful for analyzing image data (in chapter 22).

The essence of a convolution is that the intensity of each pixel in the image is replaced by a weighted sum of the neighborhood of that pixel. In the simplest case, the weight of each neighbor in the neighborhood is the same number, but things get a lot more interesting when the weights are not equal for all neighbors. The function that describes the weights that are applied to each neighbor in the neighborhood around the pixel is sometimes called a *filter* or also a *kernel*.

Let's make this explanation a bit more concrete by looking at a simple example. Because images are arrays, we can create a small array and work with it before interacting with more complicated images. To make things very concrete, we will use a function that we have created that takes an array and displays it (just like `imshow` or `matshow`), but also adds a numeral that displays the value of the pixel in that location:

```
from ndslib.viz import imshow_with_annot
import numpy as np
```

Let's start with a very small array of 4 × 5 pixels, which will serve as our example image for now. Because the pixels do not have three RGB values in them, `imshow` displays them as values of gray, where lower values are darker and the higher values are brighter (image appears on the next page):

```
small_image = np.concatenate([np.arange(10), np.arange(10, 0, -1)]).reshape((4, 5))
imshow_with_annot(small_image)
```

A convolution takes this image as input and produces an output with a similar size, but different values. To demonstrate the convolution filter operation we will start by creating

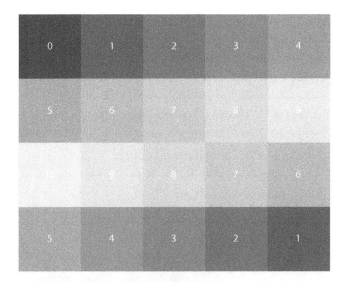

array that will eventually hold the results of this filter operation. Then, we will start filling it up with the results of the convolution as we go along:

```
small_result = np.zeros(small_image.shape)
```

The values that will eventually be placed in this result array are based on the sum of a little neighborhood around each pixel weighted by the kernel function. For our example, let's design a little 3 × 3 pixel kernel with values in it:

```
small_kernel = np.array([[0, 1, 0], [1, 1, 1], [0, 1, 0]])
imshow_with_annot(small_kernel)
```

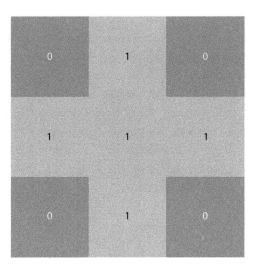

Before we can get to convolving the kernel and the image ("convolving" is the verb form of "convolution"), we have to deal with the fact that not all of the pixels have the same number of neighbors surrounding them. This is because some of the pixels are right on the edge of the image. One way to deal with that is to pad the image all around with zeros 14.5:

```
padded_small_image = np.pad(small_image, 1)
```

This adds another layer of pixels all around the image.

```
imshow_with_annot(padded_small_image)
```

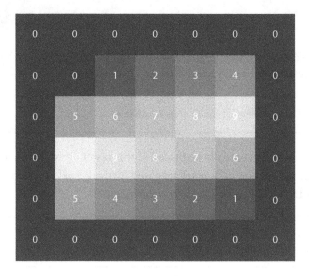

Having done that, we can start doing our convolution. We start with the pixel that is in the top left part of the image. Having added a pad of one pixel all around the image, the pixel that used to be in coordinate [0, 0] is now in coordinate [1, 1] instead. To produce a weighted sum of its neighbors, we first take the subarray small_image[0:3, 0:3] and multiply it by the kernel. Recall from chapter 8 that the multiplication of two NumPy arrays of the same size (here three elements by three elements) with the * operator produces a result of the same size, where each element is the product of the elements in that location in the original arrays. In other words, this operator produces an element-by-element product of the two arrays.

```
neighborhood = padded_small_image[:3, :3]
weighted_neighborhood = neighborhood * small_kernel
imshow_with_annot(neighborhood)
imshow_with_annot(weighted_neighborhood)
```

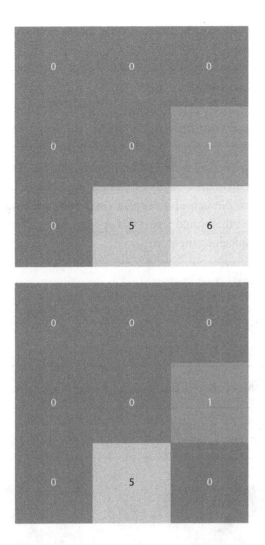

To finish the convolution operation, we want the weighted sum, so we take the values of this `weighted_neighborhood` and sum them up:

```
conv_pixel = np.sum(weighted_neighborhood)
```

This will be the array that will be placed in the top left corner of the result. We have not zero-padded the result, so that pixel is still in coordinate `[0, 0]` in the result array:

```
small_result[0, 0] = conv_pixel
```

We can repeat this same sequence of operations for the next pixel: the second pixel in the top row of the image.

```
neighborhood = small_image[:3, 1:4]
weighted_neighborhood = neighborhood * small_kernel
conv_pixel = np.sum(weighted_neighborhood)
small_result[0, 1] = conv_pixel
```

To complete the convolution, we would keep on moving like this pixel by pixel in the first row and then move on to the second row, completing each pixel in the second row and so on, until we have gone through all of the pixels in the original image. We can generalize this to write a bit of code that would repeat this operation over every pixel in the original image, filling up the result along the way:

```
for ii in range(small_result.shape[0]):
    for jj in range(small_result.shape[1]):
        neighborhood = padded_small_image[ii:ii+3, jj:jj+3]
        weighted_neighborhood = neighborhood * small_kernel
        conv_pixel = np.sum(weighted_neighborhood)
        small_result[ii, jj] = conv_pixel
```

After completing this operation, the result is an array of the same size.

```
imshow_with_annot(small_image)
imshow_with_annot(small_result)
```

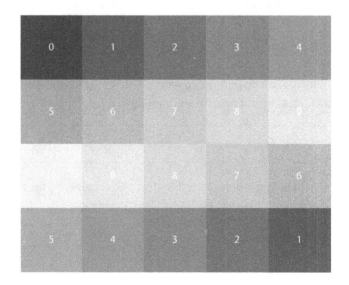

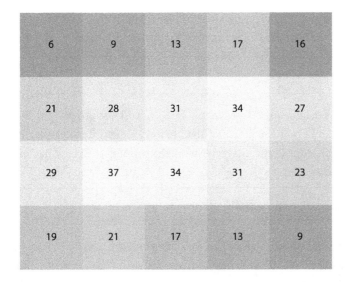

We can expand this idea to a slightly more realistic example, based on the astronaut image that you saw above. One filter that is commonly used in image processing is a Gaussian blur filter. Here, the kernel is a small Gaussian bump: in each pixel of the original image, we add in a bit of the value of the pixels in its neighborhood. The further the neighbor is, the less of that neighbor's value gets added in. This is useful if you need to blur the image a bit. For example, if you want to pool over neighboring pixels to overcome local noise patterns. We have implemented a function that generates a Gaussian kernel:

```
from ndslib.image import gaussian_kernel

kernel = gaussian_kernel()

fig, ax = plt.subplots()
im = ax.imshow(kernel, cmap="gray")
```

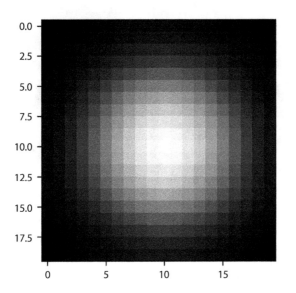

As before, we begin by creating what will be the final result, an empty array with the same size as the array of the image that we are interested in convolving with our kernel.

```
result = np.zeros(gray_img.shape)
```

Because the kernel is a bit larger in this case, we will need to pad the image around its edges with a margin that is a bit larger than before. To get the kernel right on top of the pixel that is on the edge of the image, we need the padding to extend half of the size of the kernel in each direction:

```
padded_gray_img = np.pad(gray_img, int(kernel.shape[0] / 2))
```

Once we do that, the code looks remarkably similar to the code that you saw before, the main difference being that the neighborhood in each round is now defined to be larger, extending to the size of the kernel.

```
for ii in range(result.shape[0]):
    for jj in range(result.shape[1]):
        neighborhood = padded_gray_img[ii:ii+kernel.shape[0], jj:jj+kernel.shape[1]]
        weighted_neighborhood = neighborhood * kernel
        conv_pixel = np.sum(weighted_neighborhood)
        result[ii, jj] = conv_pixel
```

The outcome is again an image with a size that is identical to the size of the original input image. In this case, it is instructive to compare the original image with the filtered image:

```
fig, ax = plt.subplots(1, 2)
ax[0].imshow(gray_img, cmap="gray")
im = ax[1].imshow(result, cmap="gray")
```

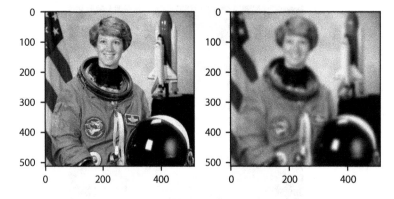

As you can see, the image on the right (filtered) is blurrier than the image on the left (the original). Intuitively, this is because each pixel is now more similar to the pixels that are its neighbors. For example, the boundaries between parts of the image that are very dark (e.g., the parts of the image that contain Commander Collins' eye) and bright parts (the skin surrounding the eye) have become fuzzier, and this transition is now more gradual. Each pixel resembles its neighbors more because they have been mixed through the process of convolution.

Exercise

MRI data of brains contain three-dimensional images of brains. Write the code that you would need to convolve an MRI data set with a three-dimensional kernel.

14.4.3 Convolution Filters in Scikit Image

Scikit Image implements a variety of standard convolutional filters. Using the implementation in Scikit Image is generally easier because it saves you the need to keep track of issues like padding and indexing into the array. In cases where the code can be speeded up using various tricks, these have been implemented so that your code will also run faster. In other words, if Scikit Image implements it, it would be a waste of your time to re-implement it. For example:

```python
from skimage.filters import gaussian
fig, ax = plt.subplots()
im = ax.imshow(gaussian(gray_img, sigma=5))
```

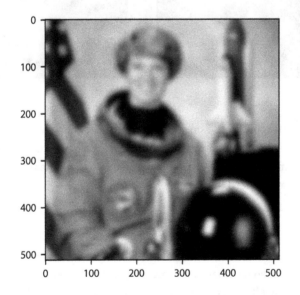

The sigma parameter that we input into the function changes the spread of the Gaussian filter that is used. To get a feel for what this parameter does, you can try changing its value to larger or smaller numbers.

One of the advantages of many of the filters implemented in Scikit Image is that they can operate on more than the two dimensions of a photograph. For example, the same function works just as well on three-dimensional scientific data, such as brain images. That means that the work that you did in the previous exercise is done for you within the call to this function. To demonstrate that, let's look at a single volume from a functional MRI (fMRI) blood oxygenation level dependent (BOLD) signal series:

```
from ndslib.data import load_data
brain = load_data("bold_volume")
```

We apply the same Gaussian filter to this data and then compare the original data with the Gaussian-smoothed data.

```
smoothed_brain = gaussian(brain, sigma=1)
fig, ax = plt.subplots(1, 2)
ax[0].matshow(smoothed_brain[:, :, 10])
im = ax[1].matshow(brain[:, :, 10])
```

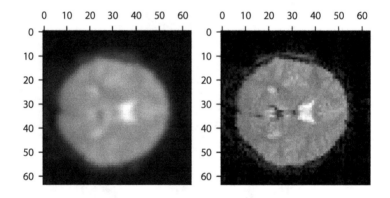

Notice that we are applying the smoothing to the three-dimensional volume of data, so each voxel receives some information not only from the voxels surrounding it within the plane of the two-dimensional image we see here but also from its neighbors in the other slices immediately adjacent to this slice.

Exercise

Scikit Image implements many other filters. The effect of the filters is not always to smooth the data. In some cases, filters are designed to emphasize certain features of the data. One of the filters that emphasize features is called a Sobel filter. Try running `skimage.filters.sobel` on the two images that you have seen here: the grayscale image of Commander Collins and the brain volume. What is the Sobel

> filter emphasizing? Try running `skimage.filters.sobel_h` and `skimage` `.filters.sobel_v` on the two-dimensional grayscale image. What do these filters emphasize? What do you think "h" and "v" stand for? Why wouldn't these functions work on the brain image?

14.4.4   Morphological Operations

Another set of filtering operations—operations that take images as their input and produce images as their outputs—pay specific attention to the shapes that appear in the image. These are known as *morphological operations* (the word morphology is derived from the Greek word for shape). Instead of a kernel, morphological operations use a *structuring element*. These are also small arrays that are used to process each pixel and its neighborhood, but there are a few ways in which structuring elements differ from kernels: first, they are binary, containing only ones and zeros. Instead of multiplying the structuring element with the region surrounding each pixel, in morphological operations, the structuring element is compared to the contents of this neighborhood in a variety of different ways. For example, in a morphological *dilation*, each pixel in the image gets the maximal value of the pixels within the regions covered by ones in the structuring element. Conversely, in morphological *erosion* each pixel gets the minimal value of the pixels covered by ones in the structuring element. Per default, the structuring element that is used in Scikit Image is exactly the kernel that we used to demonstrate convolution in Section 14.4.2: a cross-shaped structure with ones along the horizontal and vertical pixels and zeros in the corners. It turns out that this is a good structuring element to perform useful and interesting image processing operations even just with these two operations. Let's look at an example of this. For now, we will look at this using a toy example, and we will come back to look at its use with brain data in chapter 15. The Shepp-Logan phantom is an image that was created to emulate the appearance of a scan of a human brain, as the basis for developing image processing algorithms for brain imaging methods. It contains a 400 × 400 pixel array of grayscale values that include a skull (brighter than other parts) and ventricles (darker than other parts), as well as some internal structure (white matter, gray matter, and subcortical nuclei).

```
shepp_logan = skimage.data.shepp_logan_phantom()
fig, ax = plt.subplots()
im = ax.matshow(shepp_logan)
```

One of the things that erosion operations do rather effectively is take care of certain kinds of noise in the image. One example is "salt-and-pepper noise" that sometimes appears in images and takes the form of pixels that are randomly flipped to completely black or completely white. Let's add some of this kind of noise to the image:

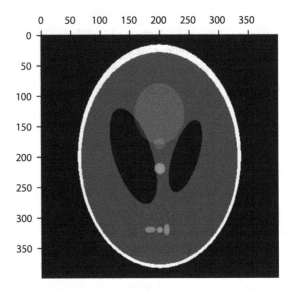

```
noise = np.random.rand(400, 400) < 0.1
fig, ax = plt.subplots()
im = ax.matshow(shepp_logan + noise)
```

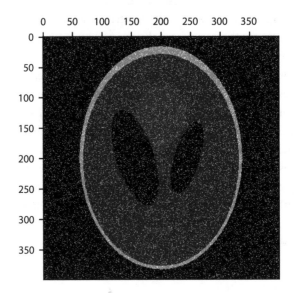

An erosion operation can be used to remove this noise from the image because the noise pixels are usually surrounded by pixels that have the correct value.

```
from skimage.morphology import erosion
fig, ax = plt.subplots()
im = ax.matshow(erosion(shepp_logan + noise))
```

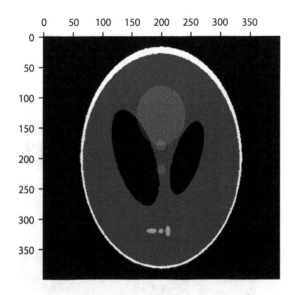

This effectively removes the noise, but it also thins out the skull-like portion of the image and the smaller elements within the image, while also slightly expanding the size of the dark ventricles. Let's correct that by applying a dilation to the eroded image. This recovers the image almost precisely as it was.

```
from skimage.morphology import dilation
fig, ax = plt.subplots()
im = ax.matshow(dilation(erosion(shepp_logan + noise)))
```

But the fact that the erosion removes a layer of pixels off of the boundaries between different segments of the image can also be useful. For example, the subtraction of the erosion from the original image can be used to detect the edges between parts of the image.

```
fig, ax = plt.subplots()
im = ax.matshow(shepp_logan - erosion(shepp_logan))
```

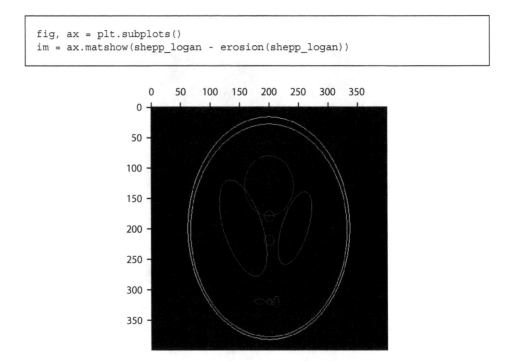

An erosion followed by a dilation is also called a *morphological opening*. This can be used to remove most of the skull-like shape from this image, provided we use a large enough structuring element. Here, we use a disk with a radius of 7 pixels. Notice that this effectively removes the skull, while retaining the original size of the ventricles, but some of the smaller elements in the image are also eliminated along the way.

```
from skimage.morphology import disk
fig, ax = plt.subplots()
im = ax.matshow(dilation(erosion(shepp_logan, selem=disk(7)), selem=disk(7)))
```

A *morphological closing* is the exact opposite: a dilation followed by an erosion. This tends to close the gaps between smaller elements in the image, which can be useful in creating boundaries between different domains represented in the image.

```
fig, ax = plt.subplots()
im = ax.matshow(erosion(dilation(shepp_logan, selem=disk(7)), selem=disk(7)))
```

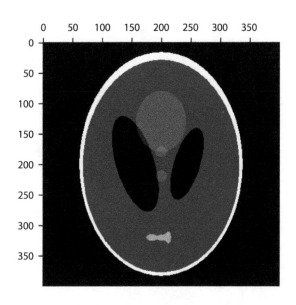

Exercises

1. A *morphological white top hat* is an image minus its morphological opening. A *morphological black top hat* is an image minus its morphological closing. Implement each of these as a function. What kinds of features do each of these emphasize? How does that depend on the structuring element that you use? Hint: `skimage.morphology` implements several structuring elements, including `disk` that you have already seen, but also `square`, `diamond` and others. Please consult the Scikit Image documentation[3] for more information.

2. Apply the morphological operations that you have learned about in this chapter to the brain slice image that we used to demonstrate Gaussian filtering. Can you use these operations to remove the bits of skull that are apparent at the top of this slice? What do you lose when you do that? Can you do better by performing your morphological operations in three dimensions? (Note: you will need to use a three-dimensional structuring element; try `skimage.morphology.ball`).

3. https://scikit-image.org/docs/dev/auto_examples/numpy_operations/plot_structuring_elements.html

14.5   Additional Resources

The Scikit Image documentation contains detailed examples of many of the image processing operations that we discussed previously, as well as many others. All of these examples can be executed directly from code that is provided in the examples gallery.[4]

---

4. https://scikit-image.org/docs/stable/auto_examples/

15

Image Segmentation

The general scope of image processing and some basic terminology were introduced in chapter 14. We will now dive deeper into specific image processing methods in a bit more detail. We are going to look at computational tasks that are commonly used in neuroimaging data analysis and we are going to demonstrate some solutions to these problems using relatively simple computations.

The first is image segmentation. Image segmentation divides the pixels or voxels of an image into different groups. For example, you might divide a neuroimaging data set into parts that contain the brain itself and parts that are non-brain (e.g., background, skull). Generally speaking, segmenting an image allows us to know where different objects are in an image and allows us to separately analyze parts of the image that contain particular objects of interest. Let's look at a specific example. We will start by loading a neuroimaging data set that has functional MRI (fMRI) blood oxygenation level dependent (BOLD) images in it:

```
from ndslib import load_data
brain = load_data("bold_volume")
```

For simplicity's sake, we will look only at one slice of the data for now. Let's take a horizontal slice roughly in the middle of the brain. Visualizing the image can help us understand the challenges that segmentation helps us tackle (image on next page).

```
slice10 = brain[:, :, 10]
import matplotlib.pyplot as plt
fig, ax = plt.subplots()
im = ax.imshow(slice10, cmap="bone")
```

First of all, a large part of the image is in the background, i.e., the air around the subject's head. If we are going to do statistical computations on the data in each of the voxels, we would like to avoid looking at these voxels. That would be a waste of time. In addition, there are parts of the image that are not clearly in the background but contain parts of the subject's head or parts of the brain that we are not interested in. For example, we do not

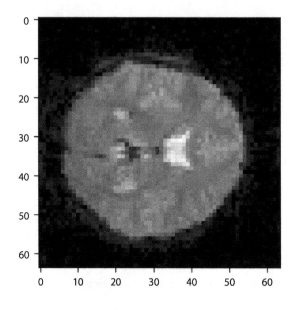

want to analyze the voxels that contain parts of the subject's skull and/or skin that appear as bright lines alongside the top of the part of the image. We also do not want to analyze the voxels that are in the ventricles, appearing here as brighter parts of the image. To be able to select only the voxels that are in the brain proper, we need to segment the image into brain and non-brain. There are a few different approaches to this problem.

15.1   Intensity-Based Segmentation

The first and simplest approach to segmentation is to use the distribution of pixel intensities as a basis for segmentation. As you can see, the parts of the image that contain the brain are brighter and have higher intensity values. The parts of the image that contain the background are dark and contain low-intensity values. One way of looking at the intensity values in the image is using a histogram. This code displays a histogram of pixel intensity values. I am using the Matplotlib `hist` function, which expects a one-dimensional array as input, so the input is the flat representation of the image array, which unfolds the two-dimensional array into a one-dimensional array (image on next page).

```
fig, ax = plt.subplots()
p = ax.hist(slice10.flat)
```

The histogram has two peaks. A peak at lower pixel intensity values, close to zero, corresponds to the dark part of the image that contains the background, and a peak at higher intensity values, corresponds to the bright part of the image that contains the brain.

One approach to segmentation would be to find a threshold value. Pixels with values below this threshold would be assigned to the background, and values equal to or above this threshold would be assigned to the brain.

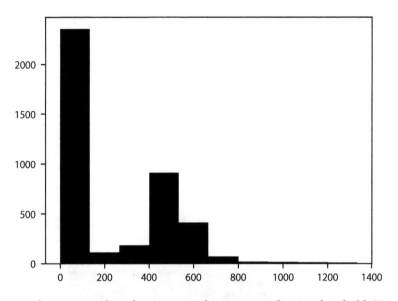

For example, we can select the mean pixel intensity to be our threshold. Here, I am displaying the image intensity histogram, together with the mean value, represented as a vertical dashed line. What would a segmentation based on this value look like?

```
import numpy as np
mean = np.mean(slice10)
fig, ax = plt.subplots()
ax.hist(slice10.flat)
p = ax.axvline(mean, linestyle='dashed')
```

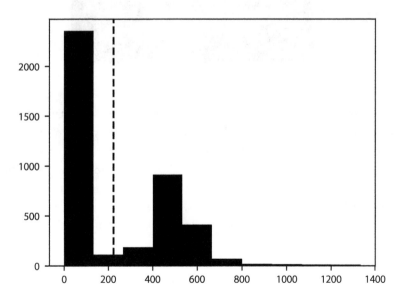

We start by creating a segmentation image that contains all zeros. Using Boolean indexing, we select those parts of the segmentation image that correspond to parts of the brain image that have intensities higher than the threshold and set these to one (see also Section 8.2.8). Then, we can display this image on top of our brain image. To make sure that we can see the brain image through the segmentation image, we set it to be slightly transparent. This is done by setting an imaging parameter called alpha, which describes the opacity of an image (where 0 is fully transparent and 1.0 is fully opaque) to 0.5. This lets us see how well the segmentation overlaps with the brain image.

```
segmentation = np.zeros_like(slice10)
segmentation[slice10 > mean] = 1

fig, ax = plt.subplots()
ax.imshow(slice10, cmap="bone")
im = ax.imshow(segmentation, alpha=0.5)
```

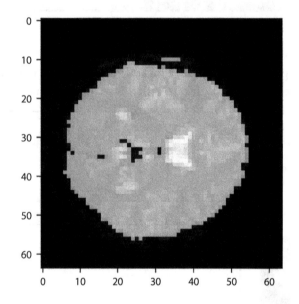

With this visualization, we can see the brain through the transparency of the segmentation mask. We can see that the mean threshold is already pretty effective at segmenting out parts of the image that contain the background, but there are still parts of the image that should belong to the background and are still considered to be part of the brain. In particular, parts of the skull and the dura mater get associated with the brain, because they appear bright in the image. Also, the ventricles, which appear brighter in this image, are classified as part of the brain. The challenge is to find a value of the threshold that gives us a better segmentation.

15.1.1   Otsu's Method

A classic approach to this problem is now known as Otsu's method, after the Japanese engineer Nobuyuki Otsu who invented it in the 1970s [Otsu 1979]. The method relies on a straightforward principle: find a threshold that minimizes the variance in pixel intensities within each class of pixels (e.g., brain and non-brain). This principle is based on the idea that pixels within each of these segments should be as similar as possible to each other, and as different as possible from the other segment. It also turns out that this is a very effective strategy for many other cases where you would like to separate the background from the foreground (e.g., text on a page).

Let's examine this method in more detail. We are looking for a threshold that would minimize the total variance in pixel intensities within each class, or *intraclass variance*. This has two components. The first is the variance of pixel intensities in the background pixels, weighted by the proportion of the image that is in the background. The other is the variance of pixel intensities in the foreground, weighted by the proportion of pixels belonging to the foreground. To find this threshold value, Otsu's method relies on the following procedure: Calculate the intraclass variance for every possible value of the threshold and find the candidate threshold that corresponds to the minimal intraclass variance. We will look at an example with some code subsequently, but we will first describe this approach in even more detail with words.

We want to find a threshold corresponding to as small as possible intraclass variance, so we start by initializing our guess of the intraclass variance to the largest possible number: infinity. We are certain that we will find a value of the threshold that will have a smaller intraclass variance than that. Then, we consider each possible pixel value that could be the threshold. In this case, that is every unique pixel value in the image (which we find using `np.unique`). As background, we select the pixel values that have values lower than the threshold. As foreground, we select the values that have values equal to or higher than the threshold.

Then, the foreground contribution to the intraclass variance is the variance of the intensities among the foreground pixels (`np.var(foreground)`), multiplied by the number of foreground pixels (`len(foreground)`). The background contribution to the intraclass variance is the variance of the intensities in the background pixels, multiplied by the number of background pixels (with very similar code for each of these). The intraclass variance is the sum of these two. If this value is smaller than the previously found best intraclass variance, we set this to be the new best intraclass variance, and replace our previously found threshold with this one. After running through all the candidates, the value stored in the threshold variable will be the value of the candidate that corresponds to the smallest possible intraclass variance.

This is the value that you would use to segment the image. Let's look at the implementation of this algorithm:

```
min_intraclass_variance = np.inf

for candidate in np.unique(slice10):
    background = slice10[slice10 < candidate]
    foreground = slice10[slice10 >= candidate]
    if len(foreground) and len(background):
        foreground_variance = np.var(foreground) * len(foreground)
        background_variance = np.var(background) * len(background)
        intraclass_variance = foreground_variance + background_variance
```

```
if intraclass_variance < min_intraclass_variance:
    min_intraclass_variance = intraclass_variance
    threshold = candidate
```

Having run through this, let's see if we found something different from our previous guess, i.e., the mean of the pixel values. We add this value to the histogram of pixel values as a dotted line. It looks like the value that Otsu's method finds is a bit higher than the mean pixel value.

```
mean = np.mean(slice10)
fig, ax = plt.subplots()
ax.hist(slice10.flat)
ax.axvline(mean, linestyle='dashed')
p = ax.axvline(threshold, linestyle='dotted')
```

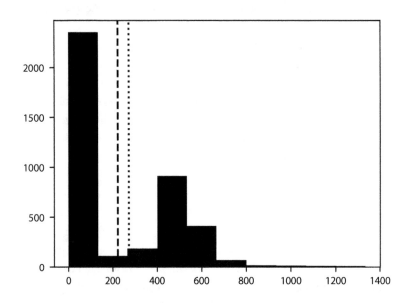

How well does this segmentation strategy work?

We will use the same code we used before, creating an array of zeros, and replacing the zeros with ones only in those parts of the array that correspond to pixels in the image that have values larger than the threshold (image on the next page).

```
segmentation = np.zeros_like(slice10)
segmentation[slice10 > threshold] = 1

fig, ax = plt.subplots()
ax.imshow(slice10, cmap="bone")
p = ax.imshow(segmentation, alpha=0.5)
```

As you can see, this segmentation is quite good. There are almost no more voxels outside the brain that are misclassified as being in the brain. We still have trouble with

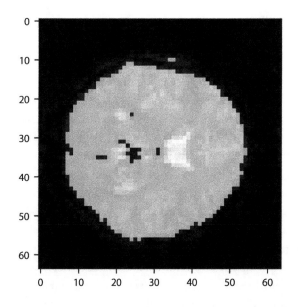

the ventricles, though. This highlights the limitation of this method. In particular, a single threshold may not be enough. In the exercise that follows you will be asked to think about this problem more, but before we move on and talk about other approaches to segmentation, we will just point out that if you want to use Otsu's method for anything you should probably use the Scikit Image implementation of it, which is in `skimage.filters.threshold_otsu`. It is much faster and more robust than the simple implementation we outlined previously. If you want to learn more about how to speed up slow code, reading through the source code of this function [1] is an interesting case study.

Exercises

1. Using the code that we implemented in the previous example, implement code that finds two threshold values to segment the brain slice into three different segments. Does this adequately separate the ventricles from other brain tissue based on their intensity? Compare the performance of your implementation to the results of Scikit Image's `skimage.filters.threshold_multiotsu`.

2. The `skimage.filters` module has a function called `try_all_threshold` that compares several different approaches to threshold-based segmentation. Run this function to compare the methods. Which one works best on this brain image? How would you evaluate that objectively?

---

1. Available at https://github.com/scikit-image/scikit-image/blob/v0.19.3/skimage/filters/thresholding.py#L312

Image Processing, Computer Vision and Neuroscience

There has always been an interesting link between image processing and neuroscience. This is because one of the goals of computational image processing is to create algorithms that mimic operations that our eyes and brain can easily do. For example, the segmentation task that we are trying to automate in this chapter is one that our visual system can learn to do with relative ease. One of the founders of contemporary computational neuroscience was David Marr. One of his main contributions to computational neuroscience was in breaking apart problems in neuroscience conceptually to three levels: computational, algorithmic, and implementational. The computational level can be thought of as the purpose of the system, or the problem that the system is solving. The algorithmic level is a series of steps or operations, ideally defined mathematically, that the system can take to solve the problem. Finally, the implementation level is the specific way in which the algorithm is instantiated in the circuitry of the nervous system. This view of neuroscience has been hugely influential and has inspired efforts to understand the nervous system based on predictions from both the computational and algorithmic levels, as well as efforts to develop algorithms that are inspired by the architecture of the brain. For example, in 1980, Marr and Ellen Hildreth wrote a paper that described the theory of edge detection and also proposed an algorithm that finds edges in images [Marr and Hildreth 1980]. Their premise was that the computational task undertaken by the first few stages in the visual system is to construct a "rich but primitive description of the image." What they called the "raw primal sketch" should provide the rest of the visual system—the parts that figure out what objects are out in the world and what they are doing—with information about things like edges, surfaces, and reflectances. Based on this computational requirement, they designed an algorithm that would specialize in finding edges. This was a particular kind of filter/kernel that, when convolved with images, would find the parts of the image where edges exist. Based on some theoretical considerations, they defined this kernel to be a difference of two Gaussians with two different variances. This edge detector was inspired to some degree by the operations that Marr and Hildreth knew the early visual cortex performs. In addition, their article proposed some predictions for what operations the visual system should be doing if this is indeed the way that it performs edge detection.

15.2   Edge-Based Segmentation

Threshold-based segmentation assumes different parts of the image that belong to different segments should have different distributions of pixel values. In many cases that is true. But this approach does not take advantage of the spatial layout of the image. Another approach to segmentation uses the fact that parts of the image that should belong to different segments are usually separated by edges. What is an edge? Usually, these are contiguous boundaries between parts of the image that look different from each other. These can be differences in intensity, but also differences in texture, pattern, and so on. In a sense, finding edges and segmenting the image are a bit of a chicken and egg dilemma: if you knew where different parts of the image were, you would not have to segment them. Nevertheless, in practice, using algorithms that specifically focus on finding edges in the image can be useful to perform the segmentation. Finding edges in images has been an important part of computer vision for about as long as there has been a field called computer vision. As you saw in chapter 14, one part of finding edges could be to construct a filter that emphasizes changes in the intensity of the pixels (such as the Sobel filter that you explored in the exercise in Section 14.4.3). Sophisticated edge detection algorithms can use this, together with a few more steps, to more robustly find edges that extend in space. One such algorithm is the Canny edge detector, named after its inventor, John Canny, a computer scientist and computer vision researcher [Canny 1986]. The algorithm takes several steps that include smoothing the image to get rid of noisy or small parts, finding the local gradients of the image (e.g., with the Sobel filter) and then thresholding the image, selecting edges that are connected to other strong edges, and suppressing edges that are connected to weaker edges. We will not implement this algorithm step by step, but instead use the algorithm as it is implemented in Scikit Image. If you are interested in learning more about the algorithm and would like an extra challenge, we would recommend also reading through the source code of the Scikit Image implementation.[2] It is just a few dozen lines of code, and understanding each step the code is taking will help you understand this method much better. For now, we will use this as a black box and see whether it helps us with the segmentation problem. We start by running our image through the Canny edge detector. This creates an image of the edges, containing a one in locations that are considered to be at an edge and zero in all other locations (image on the next page).

```
from skimage.feature import canny
edges = canny(slice10)

fig, ax = plt.subplots()
ax.imshow(slice10, cmap="gray")
im = ax.imshow(edges, alpha=0.5)
```

As you can see, the algorithm finds edges both around the perimeter of the brain as well as within the brain. This just shows us that the Canny algorithm is very good at overcoming one of the main challenges of edge detection, which is that edges are defined by intensity gradients across a very large range of different intensities. That is, an edge can be a large difference between very low pixel intensities (e.g., between the background and the edge of the brain) as well as between very high pixel intensities (parts of the brain and the ventricles). Next, we will use some morphological operations to fill parts of the image that are enclosed within edges and to remove smaller objects that are spurious (see Section 14.4.4).

---

2. Available at https://github.com/scikit-image/scikit-image//blob/v0.19.3/skimage//feature/_canny .py#L205

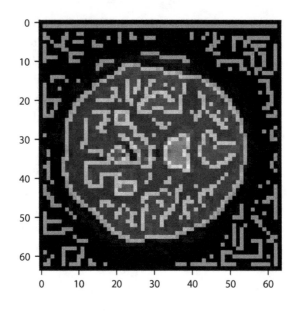

```
from scipy.ndimage import binary_fill_holes
from skimage.morphology import erosion, remove_small_objects
segmentation = remove_small_objects(erosion(binary_fill_holes(edges)))
fig, ax = plt.subplots()
ax.imshow(slice10, cmap="bone")
im = ax.imshow(segmentation, alpha=0.5)
```

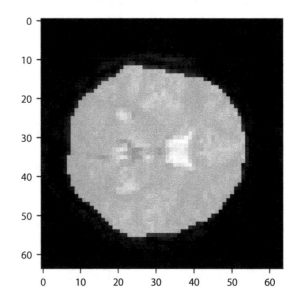

This approach very effectively gets rid of the parts of the skull around the brain, but it does not separate the ventricles from the rest of the brain. It also does not necessarily work as well in other slices of this brain (as an exercise, try running this sequence of commands on other slices. When does it work well? When does it fail?). Brain data segmentation is a very difficult task. Though we introduced a few approaches here, we do not necessarily recommend that you rely on these approaches as they are. The algorithms that are used in practice to do brain segmentations rely not only on the principles that we showed here but also on knowledge of brain structure and extensive validation experiments. Segmentation of brain data into multiple structures and tissue types, as done by the popular Freesurfer software,[3] requires such knowledge. But this does demonstrate a few of the fundamental operations that you might use if you were to start down the path of developing such algorithms for new use cases.

15.3 Additional Resources

Both Nobuyuki Otsu's original description of Otsu's method [Otsu 1979] and John Canny's original description of the Canny filter [Canny 1986] are good reads if you want to understand the use cases and development of this kind of algorithm and of image-processing pipelines more generally.

---

3. https://surfer.nmr.mgh.harvard.edu/

16

Image Registration

The last major topic we will cover in this part of the book is image registration. Image registration is something we do when we have two images that contain parts that should overlap with each other, but do not. For example, consider what happens when you take two photos of the same person. If your camera moves between the two photos, or if the person moves, their image will be in a slightly different part of the frame in each of the photos. Or, if you zoom closer between the photos, their face might take up more space in the photo. Other parts of the image that appear in the background might also be displaced. Let's look at an example of this. We will load two photos taken in close succession. In between the two photos, both the camera and the object of the photos moved. This means that the two images look a little bit different. This is apparent if we display the images side by side, but becomes even more apparent when we create a stereo image from the two images. This is done by inserting each of the images into different channels in an red-green-blue (RGB) image. In this stereo image (furthest on the right), the contents of the first photo appear in red, and the contents of the second photo appear in green. Where their contents are similar, these will blend to yellow.

```
from skimage.io import imread

naomi1 = imread('./figures/naomi1_xform.png')
naomi2 = imread('./figures/naomi2_xform.png')

import matplotlib.pyplot as plt

fig, axes = plt.subplots(1, 3, figsize=(8, 4))
ax = axes.ravel()

ax[0].imshow(naomi1)
ax[1].imshow(naomi2)

import numpy as np
stereo = np.zeros((800, 600, 3), dtype=np.uint8)
stereo[..., 0] = naomi1
stereo[..., 1] = naomi2
ax[2].imshow(stereo)

fig.tight_layout()
```

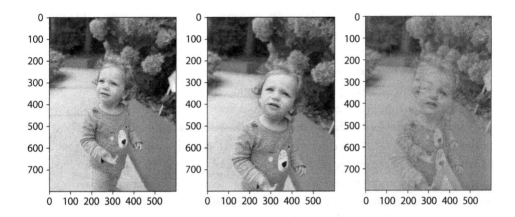

As it is, parts of the image that should overlap in the combined image are pretty far from each other (for a color version of this image, please visit https://press.princeton.edu /book/data-science-for-neuroimaging). To make these parts overlap, we have to change one of the images. This change can be a shift in the location of parts of the image, it can be a rescaling of the image, such as a zoom in or out, or it can be a rotation. As you will see later on, in some cases we will want to introduce changes to the image that are even stranger than that, to account for local changes, such as the fact that the person in these photos moved their head between the two photos. But first of all, what is this good for?

This relates to something that we discussed in chapter 12. In many experiments that we do in neuroimaging we collect images of the brain that represent different aspects of brain function and structure. For example, in a functional MRI (fMRI) experiment, we might collect images that are sensitive to the blood oxygenation level dependent (BOLD) contrast. In the same experimental session, we might also collect images that represent an anatomical contrast. For example, it is common to collect T1-weighted images that we can use to distinguish between parts of the brain that contain the cell bodies in the gray matter from nearby parts of the brain that contain the axons in the white matter. In our previous discussion of this topic in chapter 12, we showed how information is stored in the header of the NIfTI file that allows us to align two images to each other, even if the acquisition volumes differed between the two data types in terms of the voxel size, the field of view, and the orientation of the image slices. In our previous discussion, we set aside a major problem: What do we do if our subject moves between the acquisition of the BOLD fMRI data and the T1-weighted data? Or between different volumes of the same fMRI series? In these cases, image registration methods are used to calculate how we need to move images toward each other so that they maximally overlap.

16.1 Affine Registration

The first kind of registration that we will talk about corrects for changes between the images that affect all of the content in the image. These include the fact that the person in the photo is in a different part of the frame, as are details of the background like the fence above and to the right of their shoulder and the path winding to the left of them in the image. In this case, these global changes are mostly affected by the motion of the camera between the two photos. In the case of brain imaging, a global change to the position of the brain is mostly related to the fact that the subject moved between the time in which each of the images was acquired. We will see some examples with brain data further down, but let's keep going with our photographic example for now.

An affine registration can correct for global differences between images by applying a combination of transformations on the image: translating the entire image up or down, to the left or the right; rotating it clockwise or counterclockwise; scaling the image (zooming in or out); and shearing the image. While most of these are straightforward to understand, the last one is a bit more complex: *image shearing* means that different parts of the image move to different degrees. It is simplest to understand by considering a horizontal shear. In one example of a horizontal shear, the top row of pixels does not move at all, the second row moves by some amount, the third row moves by that amount, and a small amount more, and so on. A larger amount of shear means that as we go down the image subsequent rows of pixels move by more and more. For example, here are images with the amount of horizontal shear increasing from left (the original image, with no shear) to right (maximal shear):

Affine registration does relate to the affine transformation described in chapter 12. In that chapter, we used affine transformations to *align* two images, which might have different spatial resolutions, fields of view, and orientations, when a common reference frame is known to exist, i.e., the scanner space. Here, we will use an affine to register two images to each other, even without the need for a common reference frame. But, because we do not have a common reference frame, we do not know in advance what the values in the transformation need to be. Instead, we will usually try to find out what are the correct parameters for this transformation via a process of optimization. Without going into a full explanation of this process, optimization is generally the way for a program to find the right set of parameters for a particular problem. Most optimization procedures work by trying out some values of the parameters—the values of the affine transform, in this case—and then adjusting the parameters to lower the value of a *cost function*. The process continues until the algorithm finds the minimal value of this cost function. For registration, the cost function is usually how different the images are from each other when transformed using the current affine parameters.

To demonstrate this, we will use the DIPY[1] software library. Diffusion Imaging in Python (DIPY) is a software library that implements methods in computational neuroimaging. As the name suggests, it includes many methods for the analysis of

1. https://dipy.org/

diffusion MRI (dMRI) data, but it also includes methods that are useful well beyond dMRI, including methods for image registration.

We will start with a two-dimensional image registration, and apply it to these photographs. We start by importing two objects: one that represents the affine transformation in two dimensions and another that implements the registration optimization procedure.

```
from dipy.align.transforms import AffineTransform2D
from dipy.align.imaffine import AffineRegistration
```

Each of these objects is initialized and an optimization procedure is performed with the two images and the expected transform (the two-dimensions affine transform) as input, as well as a required setting for the initial parameters for the optimization (param0). Sometimes, we have some information to tell us where to initialize this optimization (e.g., based on the alignment of the two images in the scanner; see chapter 12). In this case, we do not have any information to tell us what these initial parameters should be, so we pass None.

```
affreg = AffineRegistration()
transform = AffineTransform2D()
affine = affreg.optimize(naomi1, naomi2, transform, params0=None)
```

```
Optimizing level 2 [max iter: 10000]
Optimizing level 1 [max iter: 1000]
Optimizing level 0 [max iter: 100]
```

After optimization, the optimize() method has determined the parameters of the affine, and we can apply the transform to one of the images and visualize the results in the same format that we used in the previous example.

```
naomi2_xform = affine.transform(naomi2)

fig, axes = plt.subplots(1, 3, figsize=(8, 4))
ax = axes.ravel()

ax[0].imshow(naomi1)
ax[1].imshow(naomi2_xform)

stereo = np.zeros((800, 600, 3), dtype=np.uint8)
stereo[..., 0] = naomi1
stereo[..., 1] = naomi2_xform
ax[2].imshow(stereo)
fig.tight_layout()
```

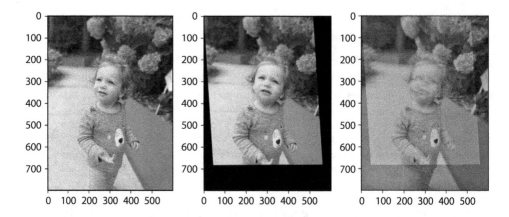

This works rather well for some parts of the image (for a color version of this image visit https://press.princeton.edu/book/data-science-for-neuroimaging). Particularly parts of the image that did not move much between the two photographs. Unfortunately, the subject of the photo also moved in a way that could not be captured by the affine transformation. This is apparent in parts of the face that could not be registered using the affine transform. Subsequently, we will see another way of registering images that can capture some of these discrepancies. But first, let's see how this applies to neuroimaging data.

16.1.1   Affine Registration of Brain Images

Another situation that requires that we *register* brain images to each other is when we want to put a specific subject's brain in register with a template. We do this when we are using an atlas that contains information that we would like to use in the individual brain, for example a brain parcellation that is available in the template space and we would like to apply it to the individual brain. Another case is when we want all of the subjects in a study to correspond to each other, i.e., when we want to summarize the data across subjects in every location.

Let's see an example of that. We will use the `templateflow` software library,[2] which gives uniform access to a variety of templates. Here, we will use the MNI152 template.[3]

```
import templateflow.api as tflow
import nibabel as nib

mni_img = nib.load(tflow.get('MNI152NLin2009cAsym', resolution=1, suffix="T1w",
   desc=None))
mni_data = mni_img.get_fdata()
```

We will also download and open a T11-weighted image from the same dataset that we previously used in chapter 12.

```
from ndslib.data import download_bids_dataset
download_bids_dataset()
t1_img = nib.load("ds001233/sub-17/ses-pre/anat/sub-17_ses-pre_T1w.nii.gz")
```

We will start by aligning the images into the same space using the same method that we used in chapter 12. In this case, we are aligning the T1-weighted image into the space of the template:

```
from nibabel.processing import resample_from_to
t1_resampled = resample_from_to(t1_img, (mni_img.shape, mni_img.affine))
t1_resamp_data = t1_resampled.get_data()
```

We can use a similar method to the one we used to compare the photos to visualize the similarity between these brain images, slicing through the three-dimensional image and visualizing just one axial slice roughly halfway through the volume. For the composite image, we are also going to have to normalize the images (dividing each one by its maximal value and multiplying by two-hundred fifty-five), so that they are roughly in the same range and blend nicely.

```
fig, axes = plt.subplots(1, 3, figsize=(8, 4))
ax = axes.ravel()

ax[0].imshow(mni_data[:, :, 85])
ax[1].imshow(t1_resamp_data[:, :, 85])

stereo = np.zeros((193, 229, 3), dtype=np.uint8)
stereo[..., 0] = 255 * mni_data[:, :, 85]/np.max(mni_data)
stereo[..., 1] = 255 * t1_resamp_data[:, :, 85]/np.max(t1_resamp_data)
ax[2].imshow(stereo)
fig.tight_layout()
```

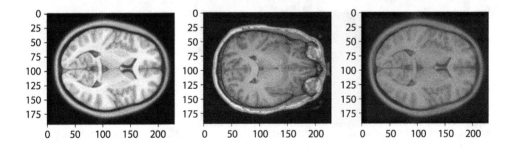

We can see that at this point, the brain is not well-registered to the template (for a color version of this image please visit https://press.princeton.edu/book/data-science -for-neuroimaging). For example, the subject's eyes in the T1-weighted image are where the frontal lobes are in the template.

We will use the same code that we used for the affine registration of the photographs before, with just one small change: instead of using the `AffineTransform2D` object, we will use the `AffineTransform3D` to register the two volumes to each other.

```
from dipy.align.transforms import AffineTransform3D
affine3d = affreg.optimize(mni_data, t1_resamp_data, AffineTransform3D(),
    params0=None)
t1_xform = affine3d.transform(t1_resamp_data)
```

```
Optimizing level 2 [max iter: 10000]
Optimizing level 1 [max iter: 1000]
Optimizing level 0 [max iter: 100]
```

```
fig, axes = plt.subplots(1, 3, figsize=(8, 4))
ax = axes.ravel()

ax[0].imshow(mni_data[:, :, 85]/np.max(mni_data))
ax[1].imshow(t1_xform[:, :, 85]/np.max(t1_xform))

stereo = np.zeros((193, 229, 3), dtype=np.uint8)
stereo[..., 0] = 255 * mni_data[:, :, 85]/np.max(mni_data)
stereo[..., 1] = 255 * t1_xform[:, :, 85]/np.max(t1_xform)
ax[2].imshow(stereo)
fig.tight_layout()
```

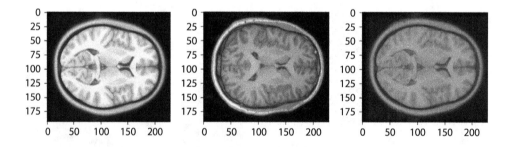

This is already not too bad (see those front lobes) and might be good enough for many applications, but you can also see that some details are still off (for a color version, please visit https://press.princeton.edu/book/data-science-for-neuroimaging). For example, the occipital horns of the lateral ventricles extend much further back in the template than in the subject's T1-weighted image. Many of the gyri and sulci of the cortex are also off. The question is: What more can we do?

16.1.2 Diffeomorphic Registration

As you saw in the results of the affine registration, this kind of global registration approach does well in registering the overall structure of one brain image to another, but it does not necessarily capture differences in small details. Another family of registration algorithms registers different parts of the image separately. In principle, you can imagine that each pixel in the first image could independently move to any location in the second image. But using this unconstrained approach, in which you can move every pixel in one image to any location in the other image, you have not registered the images to each other, you have replaced them.

Diffeomorphic registration is an approach that balances this flexibility with constraints. In principle, every pixel/voxel in the moving image could be moved to overlap with any pixel/voxel in the static image, but neighboring pixels/voxels are constrained to move by a similar amount. That is, the mapping between the moving and the static image varies smoothly in space. To demonstrate this, we will use the DIPY implementation of an algorithm that learns this kind of transformation between two images, the symmetric normalization algorithm, or SyN [Avants et al. 2008]. The application programming interface (API) for this algorithm is slightly different because you need to explicitly define the metric that the algorithm uses to figure out whether the images are similar enough to each other, as part of the optimization procedure. Here, we are going to use the cross-correlation between the images, using the CCMetric object to define this. This metric also has some other parameters that need to be defined, a smoothing kernel that is applied to the image and the size of a window of pixels over which the metric is calculated.

```
from dipy.align.imwarp import SymmetricDiffeomorphicRegistration
from dipy.align.metrics import CCMetric

metric = CCMetric(2, sigma_diff=20, radius=20)
sdr = SymmetricDiffeomorphicRegistration(metric)
```

After the diffeomorphic registration object is defined, the computation is executed using code that is very similar to what we saw for affine registration. We also pass along a pre-alignment of the two images based on the affine registration. This is a good idea, as the diffeomorphic registration is very flexible, and we want to make sure that it registers local details to each other without warping the overall shape of the objects in the image (try removing that argument and run this again to convince yourself that it is indeed a good idea).

```
mapping = sdr.optimize(naomi1, naomi2, prealign=affine.affine)
naomi2_warped = mapping.transform(naomi2)
```

In some respects, the result is not that different from the affine registration of these two images. Features that were pretty well-registered before remain pretty nicely registered, but in some places a bit of distortion has been introduced to better align some of the local features (for a color version of the comparison that follows, visit https://press.princeton .edu/book/data-science-for-neuroimaging).

```python
import matplotlib.pyplot as plt

fig, axes = plt.subplots(1, 3, figsize=(8, 4))
ax = axes.ravel()

ax[0].imshow(naomi1)
ax[1].imshow(naomi2_warped)

stereo = np.zeros((800, 600, 3), dtype=np.uint8)
stereo[..., 0] = naomi1
stereo[..., 1] = naomi2_warped
im = ax[2].imshow(stereo)
```

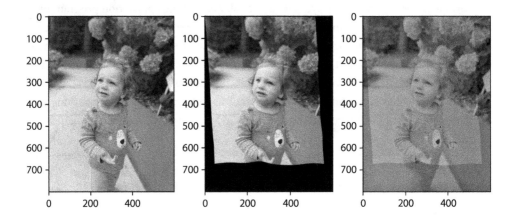

To explore this idea a bit more, we can visualize the *displacement field* between the two images, which is an image that describes how much each pixel moved as a result of the registration. In places where no displacement occurs, this looks like a regular grid. In places where there is a local distortion, the grid is distorted accordingly. This should give you a feel for the kinds of transformations that a diffeomorphism can create.

```python
from ndslib.viz import plot_diffeomorphic_map
fig, ax = plt.subplots(figsize=(4, 4))
plot_diffeomorphic_map(mapping, ax)
im = ax.imshow(naomi2_warped, alpha=0.4)
```

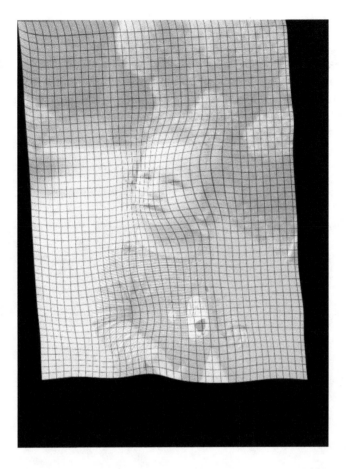

Exercise

Write code to apply the diffeomorphic registration procedure to the brain-to-template registration problem. Note that this is a three-dimensional problem, so you will need to initialize `metric = CCMetric(3)`.

Does the diffeomorphic registration resolve some of the issues we saw with affine registration? What problems do you think might arise when using diffeomorphic registration to register individual brains to a template?

16.2   Summary

Image processing algorithms are used as key steps in many neuroimaging processing pipelines. This chapter aimed to provide some intuition and the language to understand what these algorithms do, as well as a starting point to begin thinking about their use. Finally, the fundamentals of image processing provide a necessary background for understanding some of the machine learning algorithms that we will discuss in the next part of the book.

16.3   Additional Resources

The book *Elegant SciPy: The Art of Scientific Python* [Nunez-Iglesias et al. 2017] contains a chapter that explains the process of optimizing a registration between images using a cost function, with detailed code. The Jupyter notebook containing that chapter is also publicly available through the book's GitHub repository.[4]

---

4. https://github.com/elegant-scipy/notebooks/blob/master/notebooks/ch7.ipynb

PART VI

Machine Learning

17

The Core Concepts of Machine Learning

We have arrived at the final portion of the book. Hopefully you have enjoyed the journey to this point! In this chapter, we will dive into machine learning. There is a lot to cover, so to keep things manageable, we have structured the material into seven chapters. Here in chapter 17, we will review some core concepts in machine learning, setting the stage for everything that follows. In chapter 18, we will introduce the *Scikit-learn* Python package, which we will rely on it heavily throughout the chapter. Chapter 19 explores the central problem of *overfitting*; chapter 20 and chapter 21 then cover different ways of diagnosing and addressing overfitting via model validation and model selection, respectively. Finally, in chapter 22, we close with a brief review of deep learning methods, a branch of machine learning that has made many recent advances, and one that has recently made considerable inroads into neuroimaging.

Before we get into it, a quick word about our guiding philosophy. Many texts covering machine learning adopt what we might call a catalog approach: they try to cover as many of the different classes of machine learning algorithms as possible. This will not be our approach here. For one thing, there is simply no way to do justice to even a small fraction of this space within the confines of one part of a book (even a long one). More importantly, though, we think it is far more important to develop a basic grasp on core concepts and tools in machine learning than to have a cursory familiarity with many of the different algorithms out there. In our anecdotal experience, neuroimaging researchers new to machine learning are often bewildered by the sheer number of algorithms implemented in machine learning packages like Scikit-learn, and sometimes fall into the trap of systematically applying every available algorithm to their problem, in the hopes of identifying the best one. For reasons we will discuss in depth in this chapter, this kind of approach can be quite dangerous; not only does it preempt a deeper understanding of what one is doing, but, as we will see in chapters 19 and 20, it can make one's results considerably *worse* by increasing the risk of overfitting.

17.1   What *Is* Machine Learning?

This is a chapter on machine learning, so now is probably a good time to give a working definition. Here is a reasonable one: *machine learning is the field of science/engineering that seeks to build systems capable of learning from experience.*

This is a very broad definition, and in practice, the set of activities that get labeled machine learning is quite broad and varied. But two elements are common to most machine learning applications: (1) an emphasis is on developing algorithms that can learn (semi-)autonomously from data, rather than static rule-based systems that must be explicitly designed or updated by humans; and (2) an approach to performance evaluation that focuses heavily on well-defined quantitative targets.

We can contrast machine learning with traditional scientific inference, where the goal (or at least, *a* goal) is to *understand* or *explain* how a system operates.

The goals of prediction and explanation are not mutually exclusive, of course. But most people tend to favor one over the other to some extent. And, as a rough generalization, people who do machine learning tend to be more interested in figuring out how to make useful predictions than in arriving at a true, or even just an approximately correct, model of the data-generating process underlying a given phenomenon. By contrast, people interested in explanation might be willing to accept models that do not make the strongest possible predictions (or often, even good ones) so long as those models provide some insight into the mechanisms that seem to underlie the data.

We do not need to take a principled position on the prediction versus explanation divide here (plenty has been written on the topic; see additional resources at the end of this chapter). Just be aware that, for purposes of this chapter, we are going to assume that our goal is mainly to generate good predictions, and that understanding and interpretability are secondary or tertiary on our list of desiderata (though we will still say something about them now and then).

17.2   Supervised versus Unsupervised Learning

Broadly speaking, machine learning can be carved up into two forms of learning: *supervised* and *unsupervised*. We say that learning is supervised whenever we know the true values that our model is trying to predict, and hence, are in a position to supervise the learning process by quantifying prediction accuracy and the associated prediction error. Ordinary least-squares (OLS) regression, in the machine learning context, is an example of supervised learning: our model takes as its input both a vector of *features* (conventionally labeled X) and a vector of *labels* (y). Researchers often use different terminology in various biomedical disciplines—often calling X *variables* or *predictors*, and y the *outcome* or *dependent variable*—but the idea is the same.

Here are some examples of supervised learning problems (the first of which we will attempt later in the chapter):

- Predicting people's chronological age from structural brain differences
- Determining whether or not an incoming email is spam

- Predicting a person's rating of a particular movie based on their ratings of other movies
- Discriminating schizophrenics from controls based on genetic markers

In each of these cases, we expect to train our model using a data set where we know the ground truth, i.e., we have *labeled* examples of age, spam, movie ratings, and a schizophrenia diagnosis, in addition to any number of potential features we might use to try and predict each of these labels.

17.3 Supervised Learning: Classification versus Regression

Within the class of supervised learning problems, we can draw a further distinction between *classification* problems and *regression* problems. In both cases, the goal is to develop a predictive model that recovers the true labels as accurately as possible. The difference between the two lies in the nature of the labels. In classification, the labels reflect discrete classes; in regression, the labeled values vary continuously.

17.3.1 Regression

A regression problem arises any time we have a set of continuous numerical labels and we are interested in using one or more features to try and predict those labels. Any bivariate relationship can be conceptualized as a regression of one variable on the other. For example, suppose we have the data displayed in this scatterplot:

```
import numpy as np
import matplotlib.pyplot as plt
```

```
x = np.random.normal(size=30)
y = x * 0.5 + np.random.normal(size=30)

fig, ax = plt.subplots()
ax.scatter(x, y, s=50)
ax.set_xlabel('x')
label = ax.set_ylabel('y')
```

We can frame this as a regression problem by saying that our goal is to generate the best possible prediction for y given knowledge of x. There are many ways to define what constitutes the best prediction, but here we will use the *least-squares* criterion to create a model that, when given the x scores as inputs, will produce predictions for y that minimize the sum of squared deviations between the predicted scores and the true scores.

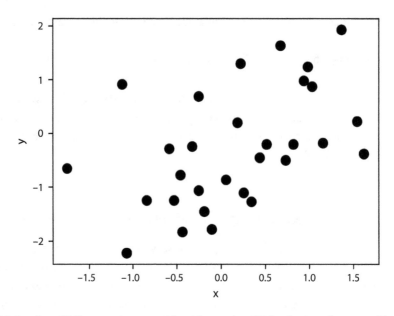

This is what OLS regression provides. To get the OLS solution, first we add a column to x. This column will be used to model the intercept of the line that relates y to x.

```
x_with_int = np.hstack((np.ones((len(x), 1)), x[:, None]))
```

Then, we solve the set of linear equations using NumPy's linear algebra routines. This gives us parameter estimates for the intercept and the slope.

```
w = np.linalg.lstsq(x_with_int, y, rcond=None)[0]
print("Parameter estimates (intercept and slope):", w)
```

```
Parameter estimates (intercept and slope): [-0.36822492  0.62140416]
```

Then we visualize the data and insert a straight line that represents the model of the data based on the regression:

```
fig, ax = plt.subplots()
ax.scatter(x, y, s=50)
ax.set_xlabel('x')
ax.set_ylabel('y')

xx = np.linspace(x.min(), x.max()).T
line = w[0] + w[1] * xx

p = plt.plot(xx, line)
```

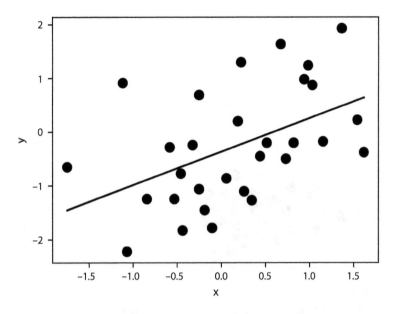

What is this model? Based on the values of the parameters, we can say that the linear prediction equation that produced the predicted scores here can be written as $\hat{y} = -0.37 + 0.62x$.

Of course, not every model we use to generate a prediction will be quite this simple. Most will not, either because they have more parameters, or because the prediction cannot be expressed as a simple weighted sum of the parameter values. But what all regression problems share in common with this simple example is the use of one or more features to try and predict labels that vary continuously.

17.3.2   Classification

Classification problems are conceptually similar to regression problems. In classification, just like in regression, we are still trying to learn to make the best predictions we can for some target set of labels. The difference is that the labels are now discrete rather than continuous. In the simplest case, the labels are binary: there are only two *classes*. For example, we can use utilities from the Scikit-learn library (we will learn more about this library starting in chapter 18) to create data that look like this:

```
from sklearn.datasets import make_blobs

X, y = make_blobs(centers=2, random_state=2)
fig, ax = plt.subplots()
s = ax.scatter(*X.T, c=y, s=60, edgecolor='k', linewidth=1)
```

Here, we have two features (on the *x*- and *y*-axes) we can use to try to correctly *classify* each sample. The two classes are labeled by color.

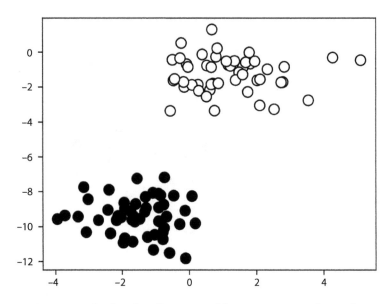

In the previous example, the classification problem is quite trivial: it is clear to the eye that the two classes are perfectly *linearly separable* so that we can correctly classify 100% of the samples just by drawing a line between them. Of course, most real-world problems will not be nearly this simple. As we will see later, when we work with real data, the feature-space distributions of our labeled cases will usually overlap considerably, so that no single feature (and often, not even all of our features collectively) will be sufficient to perfectly discriminate cases in each class from cases in other classes.

17.4   Unsupervised Learning: Clustering and Dimensionality Reduction

In unsupervised learning, we do not know the ground truth. We have a data set containing some observations that vary on some set of features X, but we are not given any set of accompanying labels y that we are supposed to try to recover using X. Instead, the goal of unsupervised learning is to find interesting or useful structure in the data. What counts as interesting or useful depends of course on the person and the context. But the key point is that there is no strictly right or wrong way to organize our samples (or if there is, we do not have access to that knowledge). So we are forced to muddle along the best we can, using only the variation in the X features to try and make sense of our data in ways that we think might be helpful to us later.

Broadly speaking, we can categorize unsupervised learning applications into two classes: clustering and dimensionality reduction.

17.4.1   Clustering

In clustering, our goal is to label the samples we have into discrete *clusters* (or groups). In a sense, clustering is just *classification without ground truth*. In classification, we are trying to recover the class assignments that we know to be there; in clustering, we are trying to make class assignments even though we have no idea what the classes truly are, or even if they exist at all.

The best-case scenario for a clustering application might look something like this:

```
X, y = make_blobs(random_state=100)
fig, ax = plt.subplots()
s = ax.scatter(*X.T)
```

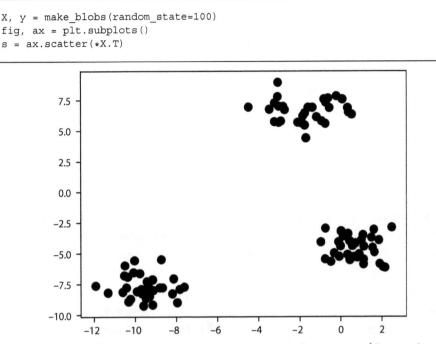

Remember: we do not know the true labels for these observations (that is why they are all assigned the same color in the previous plot). So in a sense, any cluster assignment we come up with is just our best guess as to what might be going on. Nevertheless, in this particular case, the spatial grouping of the samples in two dimensions is so striking that it is hard to imagine us having any confidence in any assignment except the following one:

```
X, y = make_blobs(random_state=100)
fig, ax = plt.subplots()
s = ax.scatter(*X.T, c=y)
```

Of course, just as with the toy classification problem we saw earlier, clustering problems this neat rarely show up in nature. Worse, in the real world, there often *are not* any true clusters. Often, the underlying data-generating process is best understood as a complex (i.e., high-dimensional) continuous function. In such cases, clustering is still helpful, as it can help reduce complexity and give us insight into regularities in the data. But when we use clustering methods (and, more generally, any kind of unsupervised learning approach), we should always try to remember the adage that *the map is not the territory*, meaning we should not mistake a description of a phenomenon for the phenomenon itself.

17.4.2 Dimensionality Reduction

The other major class of unsupervised learning application is *dimensionality reduction*. The idea, just as the name suggests, is to reduce the dimensionality of our data. The reasons why dimensionality reduction is important in machine learning will become clearer when we talk about overfitting later, but a general intuition we can build on is that most real-world data sets—especially large ones—can be efficiently described using fewer dimensions than there are nominal features in the data set. Real-world data sets tend to contain a good deal of structure: variables are related to one another in important (though often nontrivial) ways, and some variables are *redundant* with others, in the sense that they can be redescribed as functions of other variables. The idea is that if we can capture most of the variation in the features of a data set using a smaller subset of those features, we can reduce the effective dimensionality of our data set and build predictions more efficiently.

To illustrate, consider this data set:

```
x = np.random.normal(size=300)
y = x * 5 + np.random.normal(size=300)

fig, ax = plt.subplots()
s = ax.scatter(x, y)
```

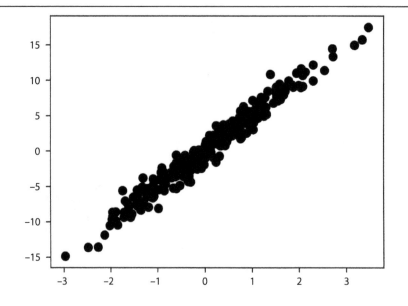

Nominally, this is a two-dimensional data set, and we are plotting the two features on the x and y axes, respectively. But it seems clear at a glance that there are not *really* two dimensions in the data—or at the very least, one dimension is far more important than the other. In this case, we could capture the vast majority of the variance along both dimensions with a single axis placed along the diagonal of the plot—in essence, rotating the axes to a simpler structure. If we keep only the first dimension in the new space and lose the second dimension, we reduce our two-dimensional data set to one dimension, with very little loss of information.

In the next section, we will dive into the nuts and bolts of machine learning in Python, by introducing the Scikit-learn machine learning library.

17.5   Additional Resources

If you are interested in diving deeper into the distinction between prediction and explanation, we really recommend Leo Breiman's classical paper "The Two Cultures of Statistical Modeling"[1] [Breiman 2001] . Another great paper on this topic is Galit Shmueli's "To Explain or to Predict?"[2] [Shmueli 2010]. Finally you can read one of us weighing in on the topic, together with Jake Westfall, in a paper titled "Choosing Prediction Over Explanation in Psychology: Lessons From Machine Learning"[3] [Yarkoni and Westfall 2017].

1. https://projecteuclid.org/download/pdf_1/euclid.ss/1009213726
2. https://projecteuclid.org/download/pdfview_1/euclid.ss/1294167961
3. https://talyarkoni.org/pdf/Yarkoni_PPS_2017.pdf

18

The Scikit-Learn Package

Now that we have a grasp on some of the key concepts, we can start *doing* machine learning in Python. In this section, we will introduce *Scikit-learn* (often abbreviated `sklearn`), which is the primary package we will be working with throughout this chapter. Scikit-learn is the most widely used machine learning package in Python, and for that matter, probably in any programming language. Its popularity stems from its simple, elegant interface, stellar documentation,[1] and comprehensive support for many of the most widely used machine learning algorithms (the main domain of Scikit-learn does not cover is deep learning, which we will discuss separately in chapter 22). Scikit-learn provides well-organized, high-quality tools for virtually all aspects of the typical machine learning workflow, including data loading and preprocessing, feature extraction and feature selection, dimensionality reduction, model selection, evaluation, and so on. We will touch on quite a few of these as we go along.

18.1   The ABIDE II Data set

To illustrate how Scikit-learn works, we are going to need some data. Scikit-learn is built on top of NumPy, so in theory, we could use NumPy's random number generation routines to create suitable arrays for Scikit-learn, just as we did earlier. But that would be kind of boring. We already understand the basics of NumPy arrays at this point, so we can be a bit more ambitious here, and try to experiment with machine learning using a real neuroimaging data set. For most of this part of the book, we will use a data set drawn from the *Autism Brain Imaging Data Exchange II* (ABIDE II) project. ABIDE is an international consortium aimed at facilitating the study of autism spectrum disorder (ASD) by publicly releasing large-scale collections of structural neuroimaging data obtained from thousands of participants at dozens of research sites [Di Martino et al. 2017]. In this chapter, we will use data from the second collection (hence the II in ABIDE II). To keep things simple, we are going to use a lightly preprocessed version of the ABIDE II data set, provided in Bethlehem et al. [2020].

---

1. https://scikit-learn.org/stable/documentation.html

```
from ndslib.data import load_data
abide_data = load_data("abide2")
```

We can get a quick sense of the data set's dimensions:

```
abide_data.shape
```

```
(1004, 1446)
```

And here are the first five rows:

```
display(abide_data.head(5))
```

```
            site  subject        age  sex  group  age_resid  fsArea_L_V1_ROI  \
0  ABIDEII-KKI_1    29293   8.893151  2.0    1.0  13.642852           2750.0
1  ABIDEII-OHSU_1   28997  12.000000  2.0    1.0  16.081732           2836.0
2   ABIDEII-GU_1    28845   8.390000  1.0    2.0  12.866264           3394.0
3  ABIDEII-NYU_1    29210   8.300000  1.0    1.0  13.698139           3382.0
4  ABIDEII-EMC_1    29894   7.772758  2.0    2.0  14.772459           3080.0

   fsArea_L_MST_ROI  fsArea_L_V6_ROI  fsArea_L_V2_ROI  ...  fsCT_R_p47r_ROI  \
0             306.0            354.0           2123.0  ...            3.362
1             186.0            354.0           2261.0  ...            2.809
2             223.0            373.0           2827.0  ...            2.435
3             266.0            422.0           2686.0  ...            3.349
4             161.0            346.0           2105.0  ...            2.428

   fsCT_R_TGv_ROI  fsCT_R_MBelt_ROI  fsCT_R_LBelt_ROI  fsCT_R_A4_ROI  \
0           2.827             2.777             2.526          3.202
1           3.539             2.944             2.769          3.530
2           3.321             2.799             2.388          3.148
3           3.344             2.694             3.030          3.258
4           2.940             2.809             2.607          3.430

   fsCT_R_STSva_ROI  fsCT_R_TE1m_ROI  fsCT_R_PI_ROI  fsCT_R_a32pr_ROI  \
0             3.024            3.354          2.629             2.699
1             3.079            3.282          2.670             2.746
2             3.125            3.116          2.891             2.940
3             2.774            3.383          2.696             3.014
4             2.752            2.645          3.111             3.219

   fsCT_R_p24_ROI
0           3.179
1           3.324
2           3.232
3           3.264
4           4.128

[5 rows x 1446 columns]
```

We can see that each row contains data for a single participant and each column represents a different variable. The first five columns contain key identifiers and phenotypic variables: respectively, these include the research site the subject comes from (our data set combines data from seventeen different sites); the unique ID of the subject, and their age (in years), sex, and diagnosis group (where one indicates an autism diagnosis and two indicates a control subject). The sixth column is not in the original data set and is a residualized version of the age variable that we will talk more about later.

The remaining 1,440 columns contain four sets of structural brain features extracted using the widely used FreeSurfer package [Fischl 2012]. Respectively, these include measures of surface area (fsArea), volume (fsVol), cortical thickness (fsCT), and local gyrification (fsLGI). Each set of FreeSurfer features contains three-hundred sixty variables, reflecting the three-hundred sixty regions of interest in the Human Connectome Project's multi modal parcellation [Glasser et al. 2016], which looks roughly like this (each color is a different region of interest, or ROI):

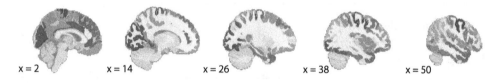

18.1.1   Data in Scikit-Learn

Now that we have a suitable data set, we need to get it to play well with Scikit-learn. It turns out we do not have to do much! As we mentioned earlier, Scikit-learn is built on top of NumPy and expects NumPy arrays as its inputs, but our abide_data variable is a Pandas DataFrame. Fortunately, Pandas DataFrame objects are built on top of the NumPy array object, so it turns out that passing our data to Scikit-learn is straightforward. The main thing we need to understand is how Scikit-learn expects its inputs to be structured, and verify that our data set respects those expectations.

FEATURE DATA: THE X MATRIX

The most important thing to know is that all model-fitting routines in Scikit-learn expect to receive a two-dimensional NumPy array—conventionally named X—as their mandatory first input. The X is expected to contain samples (i.e., independent observations) in rows and features in columns (this should remind you of the tidy data mentioned in chapter 9). For example, suppose we want to build a machine learning model that predicts a person's chronological age from their structural brain data. Then our X data will be expected to contain each participant's data on a separate row, and the columns of that row will be the values extracted for the brain features.

You will probably notice that the expected structure of X is almost the same as the structure of our ABIDE II data set. There too, we had participants in rows and features in columns. How convenient! The only thing that might concern us is that our ABIDE II data set currently contains not only the brain features extracted with FreeSurfer, but also a few identifier/phenotype columns that we probably would not want to pass to Scikit-learn as part of the X matrix (e.g., because it would not make sense to try to predict age from subject ID, and because site values are strings, whereas Scikit-learn expects X to contain only numbers). For the sake of clarity, then, let's break up our abide_data data set into two separate data frames: one containing brain variables (we will call it features), and one containing phenotypic information (`phenotypes`). This is not strictly necessary, as we can always select only a subset of columns from `abide_data` to use as X, but it will help keep our code clearer and tidier. The `DataFrame.filter` method allows us to select variables by name. In this case, we put all features that match `fs`, which stands for FreeSurfer, in the `features` DataFrame:

```
features = abide_data.filter(like='fs')
```

And remember integer-based location indexing in Pandas? Here we grab the first six columns:

```
phenotypes = abide_data.iloc[:, :6]
```

Let's verify that our `phenotypes` DataFrame now contains only the five phenotypic columns:

```
phenotypes.head(5)
```

```
            site  subject        age  sex  group   age_resid
0   ABIDEII-KKI_1    29293   8.893151  2.0    1.0   13.642852
1  ABIDEII-OHSU_1    28997  12.000000  2.0    1.0   16.081732
2    ABIDEII-GU_1    28845   8.390000  1.0    2.0   12.866264
3   ABIDEII-NYU_1    29210   8.300000  1.0    1.0   13.698139
4   ABIDEII-EMC_1    29894   7.772758  2.0    2.0   14.772459
```

LABELS: THE Y VECTOR

For unsupervised learning applications in Scikit-learn, the feature data in X are all we need. But for supervised applications (i.e., classification and regression), where we are using X to try and recover some known ground truth, the features are not enough; we also need labels, which are conventionally labeled y. Scikit-learn expects y to be a one-dimensional array (or vector).

The variables we use as labels will vary across our examples, but let's start by assigning age to y. As mentioned previously, Pandas objects are built on top of a numpy array, and we can access this array by referring to the `values` attribute of the `Series` object:

```
y = phenotypes['age'].values
```

18.2 Regression Example: Brain-Age Prediction

Now we are ready to train some machine learning models! Let's start with a regression example. We will use Scikit-learn to try to predict the measured chronological age of subjects in our data set from variations in their brain features. Predicting the biological age of a person from their neuroimaging data is a rather common use-case of machine learning. This is not just a neat party trick, though. There are reasons to think that a prediction of what is called brain age could also be scientifically useful. That is because certain conditions are associated with specific kinds of errors in brain age prediction.

18.2.1   Estimators in Scikit-Learn: Basic Usage

One of Scikit-learn's most attractive features is its simple, unified interface for configuring estimators and fitting models. A good deal of the package consists of a very large set of Estimator classes you can use to do various forms of machine learning. In Scikit-learn, an Estimator does exactly what the word *estimator* normally means in statistics: it implements a rule for estimating some parameter(s) of interest from data. In much of this chapter, we will use the terms *model* and *estimator* interchangeably, though it would be more technically accurate to say that an estimator relies internally on a particular model to generate its estimates.

While Scikit-learn includes dozens of different estimators that generate predictions in very different ways, they all share a common *application programming interface* (API)—meaning, the user interacts with them in a very similar way. In particular, every estimator class implements a fit() method. When we call an estimator's fit() method, we are telling the estimator to take some training data and do something with it. For supervised learning estimators, which we will focus on first, we have to pass two arguments to fit(): an X matrix containing the feature data we want to use to generate predictions, and a y vector containing the true labels or scores. The goal of the training process is to try to recover the labels in y from the feature data in X.

Once training is complete, we can call any supervised estimator's predict() method, which takes an X matrix as input and generates corresponding predictions for the y scores. The X matrix we provide to predict() can be the same as the one we used to fit() the estimator, though for reasons that will become clear later, doing that is often a spectacularly bad idea.

Conceptually, we can think of fit() and predict() as mapping onto distinct training and application phases: in the training phase, our model learns how to make predictions, and in the application phase we deterministically use the information our model has learned to make predictions.

18.2.2   Applying the LinearRegression() Estimator

To see how this works in practice, let's try out one particular estimator in Scikit-learn: ordinary least-squares (OLS) regression. We will start out small, using just a handful of brain features (rather than all 1,440) to generate a prediction. Let's sample five features at random from the full set we have stored in features.

We will use the fact that Pandas `DataFrame` objects have a helpful sample method for randomly sampling rows/columns. Passing a `random_state` argument allows us to pass a fixed random seed, giving us the same results each time we rerun this code.

```
n_features = 5

X = features.sample(n_features, axis=1, random_state=100)
```

Next, we import and initialize our linear regression estimator. Scikit-learn organizes estimators and other tools into modules based on what they are used for and/or the kind of model they represent. Linear models, including `LinearRegression`, are generally found in the `sklearn.linear_model` module.

```
from sklearn.linear_model import LinearRegression
```

An important principle of Scikit-learn is that the initialization of an estimator instance does not require any data to be passed. At initialization, only configuration parameters are passed. In particular, the `LinearRegression` estimator has relatively few configurable parameters and for our purposes the default settings are all fine (e.g., by default, `fit_intercept=True`, so Scikit-learn will automatically add an intercept column to our predictor matrix). This means that we initialize this object without any inputs:

```
model = LinearRegression()
```

FITTING THE MODEL

Now we are ready to fit our data! As noted previously, we do this by calling the `.fit()` method. This is true for every `Estimator` in Scikit-learn.

```
model.fit(X, y)
```

```
LinearRegression()
```

Once we execute the line shown here, we have a fitted model. One thing we can do at this point is to examine the estimated model parameters. The Scikit-learn convention is to denote fitted parameters with a trailing underscore.

```
print("Estimated intercept:", model.intercept_)
print("Estimated coefficients:", model.coef_.round(2))
```

```
Estimated intercept: 53.81302191798251
Estimated coefficients: [-3.94  0.02 -8.92 -0.   -0.01]
```

The coefficients vary dramatically in size. This is not because two of them are much more important than the other three; it is because the four sets of FreeSurfer features are on very different scales (e.g., the surface area features have much larger values than the cortical thickness features). Later in the tutorial, we will explicitly standardize the features so they are all on the same scale. But for now, it makes no difference as the predictions we get out of a `LinearRegression` estimator are insensitive to scale.

GENERATING PREDICTIONS

If we want to, we can use the parameter estimates extracted previously to explicitly compute predictions for new observations, i.e., we would effectively be applying the following prediction equation (note that we are rounding to two digits for convenience; hence, the fourth variable drops out, because its coefficient is very close to 0):

$$\hat{y} = 53.81 - 3.94x_1 + 0.02x_2 - 8.92x_3 - 0.01x_5$$

But we do not need to do this manually; we can easily generate predicted scores for new X data using the `.predict()` method that all supervised Scikit-learn estimators implement. For example, here are the predicted scores for our original X data:

```
y_pred = model.predict(X)

print(y_pred)
```

```
[15.69036168 11.17854891 16.67486452 ... 11.42650131 10.54443185
 14.57754089]
```

Let's visualize the predicted scores against the true scores. We will make use of the `jointplot` plotting function in the Seaborn library we introduced in chapter 10 (image on next page).

```
import seaborn as sns
g = sns.jointplot(x=y_pred, y=y, kind="reg").set_axis_labels("age
 ↪ (predicted)", "age (true)")
```

Two things jump out at us from this plot. First, the model appears to be capturing some signal in the data, in that the predicted age values bear some nontrivial relationship to the true values. Second, the relationship appears to be nonlinear, suggesting that our model might be misspecified. There are various ways we could potentially address this (as an exercise, try log-transforming age and repeating the estimation), but we will not worry about that here as our goal is to get familiar with machine learning in Scikit-learn, not to produce publishable results.

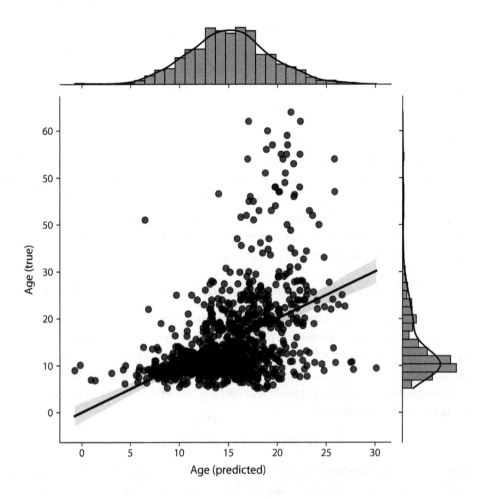

The key takeaway is that, in just a few lines of code, we have initialized a linear regression model, fitted it to some data, and generated new predicted scores. This basic pattern is common to all supervised estimators in Scikit-learn.

Just to underscore how little we had to do, here is the whole example again, in three lines:

1. Initialize the linear regression estimator.
2. Fit the model.
3. Generate predictions.

```
model = LinearRegression()
model.fit(X, y)
y_pred = model.predict(X)
```

Once we have fitted our model, we are going to want to see some results. At this point, if you are used to a point-and-click package like SPSS, or a statistics-oriented language like R, you might expect to see a big dump of information that includes things like regression coefficients, standard errors, p-values, R^2 values, and so on. Well… we are not going to talk about those things here. We *could* get that kind of information out of other Python packages (see Section 18.5). So, yes, we *could* do this kind of thing in Python. But that is not really what machine learning (or Scikit-learn) is about. Instead, the focus in machine learning is on prediction. Typically, we have some quantitative metric of predictive performance we care about, and to the degree that a model can produce better values on that metric, we are inclined to evaluate the model more favorably. This does not mean that we have to single-mindedly base our evaluation of the model on *just* one quantity; in practice, many other considerations may come into play (e.g., computational efficiency, interpretability). The point is just that machine learning practitioners tend to care much more than traditional scientists do about what models can *do*, and are much less interested in drawing conclusions about the estimated values of the model's internal parameters. With that performance-oriented goal in mind, let's spend a bit of time thinking about how to measure the quality of a model's predictions.

18.2.3 Quantifying Performance in Scikit-Learn: The Metrics Module

There are many metrics we could use to quantify the quality of the predictions our model generates. Scikit-learn conveniently packages some of the most commonly used performance metrics in its `metrics` module. As is true of `Estimator` objects, all metric functions in Scikit-learn have the same interface: we pass in the true scores and the model's predicted scores, respectively, and the function returns a number quantifying the relationship between the two.

THE COEFFICIENT OF DETERMINATION

Let's see how this works using one of the most commonly used metrics in regression problems: the coefficient of determination, or R^2, which quantifies the proportion of variance in the outcome variable (e.g., age) explained by the fitted model. We start by importing the `r2_score` function from the `sklearn.metrics` module. Then, we generate predicted values so we can compare them with the ground truth values. Generally, scoring functions are called by passing an array of true scores and an array of predicted scores as inputs, which is what we do here:

```
from sklearn.metrics import r2_score

y_pred = model.predict(X)
r2_score(y, y_pred)
```

```
0.2000318750191794
```

Our fitted linear regression model explains around 20% of the variance in age. Whether this is good or bad is a matter of perspective, but at the very least it is clear that we can nontrivially predict people's ages from a handful of structural brain features.

Exercises

1. Replace `r2_score` in the code from the previous example with `mean_squared_error`, `mean_absolute_error`, or one of the other predefined metrics in the `metrics` module.
2. Write your own metric function. To be a valid metric function it should take the true scores and predicted scores (in that order) as the only arguments and compute a single number.

BUILT-IN SCORING

For convenience, supervised Scikit-learn estimators have a `.score()` method we can use as an alternative to the previous example. Instead of generating predicted scores and then explicitly feeding them to a metric function like `r2_score`, we can call `.score()` directly on the estimator after the `fit()` step, and the prediction will be done implicitly. The only downside of this approach is that we lose the ability to specify which scoring metric to use; the choice is made for us. In the example that follows, we will initialize a linear regression, fit it with data, and then instead of generating predictions and comparing them to the data, we directly produce the score. The `LinearRegression` object uses R^2 for scoring.

```
est = LinearRegression()
est.fit(X, y)
print(est.score(X, y))
```

```
0.2000318750191794
```

18.3   Classification Example: Autism Classification

Now let's look at classification. In this case, the target labels we are trying to predict are discrete. In general, one can always turn a regression problem into a classification problem by discretizing the data in some way. For example, we could binarize our continuous age variable around the median value, giving us two equal-sized young and old groups. You will see this done a lot in the literature, but frankly, discretizing continuous data for classification purposes is almost always a really bad idea (if you are interested in learning more about that, refer to MacCallum et al. [2002]), and we are mostly mentioning that it *can* be done to point out that it generally *should not* be done.

Fortunately, no continuous variables were harmed in the making of this book. Instead of dichotomizing continuous labels, we will use a different set of labels that are naturally discrete: diagnosis group. Recall that ABIDE II is a project primarily interested in understanding autism, and roughly half of the participants in the data set are diagnosed with autism (one might reasonably argue that the pathologies *underlying* autism could be dimensional rather than discrete, but the diagnoses themselves are discrete). So let's see if we can predict autism diagnosis, rather than age, from structural brain features.

To get a very rough qualitative sense of how difficult a prediction problem this is likely to be, we can do a couple of things. First, we can look at the bivariate correlations between age and each feature in turn, sorting them by the strength of association. The .corrwith() is a DataFrame method that allows us to correlate each column with another column taken from a passed Series or DataFrame.

```
corrs = features.corrwith(phenotypes['group'])
print(corrs.sort_values().round(2))
```

```
fsCT_L_OFC_ROI        -0.14
fsCT_L_10d_ROI        -0.12
fsCT_R_31pd_ROI       -0.10
fsVol_L_10d_ROI       -0.10
fsCT_L_10v_ROI        -0.09
                       ...
fsVol_L_STSvp_ROI      0.15
fsArea_R_OFC_ROI       0.15
fsLGI_R_OFC_ROI        0.15
fsLGI_L_OFC_ROI        0.19
fsArea_L_OFC_ROI       0.19
Length: 1440, dtype: float64
```

We can immediately see that none of our 1,440 features are very strongly correlated with the diagnosis group (the largest correlations are around $r = 0.19$). However, this does not mean that all is lost; even if each feature is individually only slightly predictive of the diagnosis group individually, the full set of 1,440 features could still conceivably be very strongly predictive of the diagnosis group in the aggregate.

One way to get a cursory sense of whether *that* might be true (i.e., whether combining features is likely to help separate autistic participants from controls) is to visualize the diagnosis group as a function of a few brain features. We cannot visualize very well in more than three dimensions, so let's pick the three most strongly correlated features and use those. To probe for the separability of classes based on combinations of these variables, we plot three scatter plots of the relationships between them.

```
fs_vars = ['fsLGI_L_OFC_ROI', 'fsLGI_R_STSvp_ROI', 'fsCT_L_10d_ROI']
x, y, z = features[fs_vars].values.T

import matplotlib.pyplot as plt
fig, ax = plt.subplots(1, 3)
```

```
ax[0].scatter(x, y, c=phenotypes['group'])
ax[0].set(xlabel=fs_vars[0], ylabel=fs_vars[1])
ax[1].scatter(z, y, c=phenotypes['group'])
ax[1].set(xlabel=fs_vars[2], ylabel=fs_vars[1])
ax[2].scatter(x, z, c=phenotypes['group'])
ax[2].set(xlabel=fs_vars[0], ylabel=fs_vars[2])
fig.set_tight_layout("tight")
fig.set_size_inches([10,4])
```

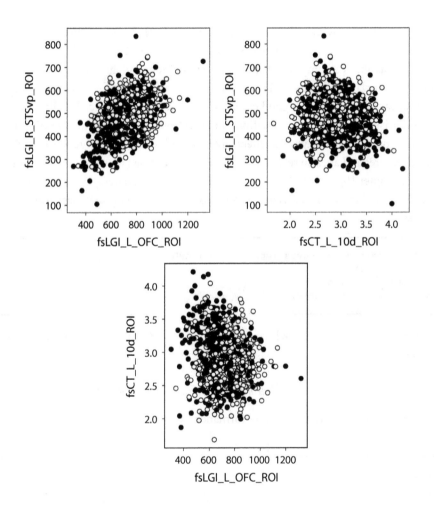

Remember in Section 17.3.2, when we mentioned that in real-world classification problems, classes are often very hard to separate cleanly, and distributions tend to overlap heavily in feature space? Well, that is exactly what we are looking at here. The black and white dots are the two groups of participants, and it is pretty clear at a glance that there is not going to be any hyperplane we could plunk down through that plot that would perfectly separate the two groups. Again though, we are still using only three of our features here, and there are 1,337 others. So let's see how we do when we scale things up to higher dimensions.

18.3.1 Applying Classifiers

Okay, onto the actual classification. How do we apply classification estimators in Scikit-learn? It is a trick question! We have already covered it in the regression example. There is essentially no difference in the way we interact with regression and classification estimators; we just have to be careful not to pass discrete labels to regression estimators, or continuous labels to classification estimators. But mechanically, we construct and fit the models in the same way.

Let's try this out with a *Gaussian Naive Bayes* (GNB) classifier. This is a simple classification approach based on a naive application of Bayes' Theorem. The naivete stems from the classifier's assumption that all features are independent of one another once we condition on class assignment. This assumption greatly simplifies analysis, though it is nearly always false. GNB is a good classifier to use as a performance baseline, because it does surprisingly well in many situations and is extremely computationally efficient, so it should be quick. Naive Bayes classifiers have no trouble handling large sets of highly correlated features and are also relatively resilient to overfitting (we will discuss overfitting in detail in chapter 19, so do not worry if the term is new to you here). So we will just throw everything we have at the classifier, and try to predict the diagnosis class from all 1,440 features.

```
from sklearn.naive_bayes import GaussianNB
gnb = GaussianNB()
y = phenotypes['group']
gnb = gnb.fit(features, y)
```

That is all there is to it! See how similar the code is to the regression example we did before?

18.3.2 Evaluating Classification Performance

We already know how to obtain a performance estimate from the built-in scorer, so let's do that:

```
print(gnb.score(features, y))
```

```
0.6254980079681275
```

In this case, the default scoring metric is just overall accuracy (i.e., the proportion of all samples that were assigned to the correct class). We see that the model classifies about 63% of samples correctly.

Is 63% good or bad? As always, it depends. It is nowhere near good enough to be useful in practical applications; conversely, it is better than randomly guessing. Or is it? That would only be the case if there is an equal number of participants that were diagnosed with autism as there are healthy controls. In some data sets, there might be a *class imbalance*, in which case, randomly guessing might do better than 50% (e.g., by guessing that every-one belongs to the more prevalent class). This is something that needs to be taken into account when evaluating classification results. Fortunately, we can compute some more nuanced scores.

CLASSIFICATION REPORTS

The raw accuracy score is a good place to start when evaluating performance, but it often masks important nuances. We can get some more information using the `classification_report` utility, which breaks down classification performance into separate `precision`, `recall`, and `f1-score` metrics (we could also get each of these individually from the `sklearn.metrics` module). Precision is also known as positive predictive value; it tells us the proportion of cases labeled as positive that truly are positive, i.e., the proportion of cases the classifier labels autistic that really are autistic. Recall (or sensitivity) tells us the proportion of true positive cases that were labeled as such by the classifier. The F1 score is the harmonic mean of recall and precision and thus tries to summarize both using a single quantity.

The `classification_report` function reports these quantities both for the data set as a whole, and broken down separately for each class:

```
from sklearn.metrics import classification_report
y_pred = gnb.predict(features)
print(classification_report(y, y_pred))
```

```
               precision    recall  f1-score   support

         1.0       0.63      0.44      0.52       463
         2.0       0.62      0.78      0.69       541

    accuracy                           0.63      1004
   macro avg       0.63      0.61      0.61      1004
weighted avg       0.63      0.63      0.61      1004
```

Notice that while the precision, in this case, is almost identical for the two groups, the recall (or sensitivity) differs substantially: control subjects have a higher probability (78%) of being correctly labeled than subjects with an autism diagnosis (44%). This suggests a bias in the model, namely that the model is more likely to assign cases to the control group than the autism group. We can verify this by looking at the overall proportion of cases our trained model labels as controls:

```
(y_pred == 2).mean()
```

```
0.6762948207171314
```

Sure enough, over two-thirds of cases are classified as controls, even though the ground truth is that only 54% of the sample is made up of controls.

Note that whether this bias in the trained model is a good or bad thing depends on one's context and goals. In the real world, there are often asymmetries in the costs associated with assigning different labels. For example, an HIV test should probably try very hard to minimize false negatives, because the cost of incorrectly telling someone they do not have HIV is likely to be higher than the cost of incorrectly telling them they *do* have HIV (seeing as follow-up tests will rapidly identify the latter error). Conversely, if all we care about is maximizing overall classification accuracy for its own sake (e.g., as a learning exercise), then we might be perfectly happy to accept a model with these kinds of class-wise biases, provided their introduction helps us improve our overall performance. We will spend much more time talking about tradeoffs and context-dependence in chapter 21, when we talk about the bias-variance tradeoff.

18.4   Clustering Example: Are There Neural Subtypes of Autism?

Recall that clustering is a form of unsupervised learning where we seek to assign our observations to discrete clusters in the absence of any knowledge of the ground truth (if there is one). Clustering applications are common in psychiatry and psychiatric imaging, as researchers often want to determine whether, e.g., patients with a particular diagnosis appear to cluster into somewhat distinct subtypes, with potential implications for prognosis, treatment, etc.

We can ask this question for the ABIDE data we have available. If we take only those subjects with an autism diagnosis, can we cluster subjects into discrete subtypes based on differences in the brain features we have available?

At this point, it probably will not surprise you to hear that Scikit-learn contains implementations of quite a few popular clustering algorithms. Clustering estimators in Scikit-learn are located, as you might intuit, in the `sklearn.cluster` module. We will focus our attention on what is arguably the most widely used clustering algorithm, namely, k-means. In k-means clustering, we assign samples to k discrete clusters in such a way as to minimize the distance from each observation to the centroid of the cluster it belongs to, and maximize the distance between the cluster centroids. We will not get into the details of the k-means algorithm here; instead, we will simply demonstrate how we would go about running a cluster analysis with Scikit-learn. As you might expect by now, our code will look a lot like it did for regression and classification.

In principle, we could use all 1,440 features if we wanted to (though computation would likely be slow). But the unsupervised nature of clustering means that evaluating clustering solutions in high dimensions can be quite difficult. So let's focus on clustering observations in just two dimensions, and then later you can crank up the number of features if you desire.

```
n_features = 2
```

Next, we will use a few lines of code that will get our X data. We start by selecting subjects. We select only the subjects with an autism diagnosis and then randomly sample brain features.

```
aut_grp = phenotypes['group'] == 1
dx_1 = features[aut_grp].sample(n_features, axis=1)
columns = dx_1.columns
```

The `sklearn.preprocessing` module contains a bunch of useful utilities. In this case we are going to standardize our columns (i.e., to transform them so they have a mean of zero and variance of one). Otherwise, k-means will likely weigh some features more than others.

```
from sklearn.preprocessing import scale
dx_1 = scale(dx_1)
```

The actual clustering is, as usual, just a couple of lines of code. We will use `KMeans` here, and we would encourage you to experiment with others, but note that many of the implemented clustering algorithms will be very slow if you crank up the number of features. Here, we have to stipulate the number of clusters (k) in advance and this is set even before fitting to the data. Instead of fitting and then predicting, we do both of these operations in one shot, using the `fit_predict` method of the `KMeans` class.

```
from sklearn.cluster import KMeans
K = 4
km = KMeans(K)
clusters = km.fit_predict(dx_1)
```

Notice that the only real difference between this code and our earlier regression/classification examples is that we are no longer passing in a y array of labels, mostly because there is not one! We are just trying to find meaningful structure in the data, with no access to any ground truth.

Also notice that we need to specify the number of clusters k ourselves. There are literally hundreds of methods people have developed to try to identify the optimal value of k, and some of them are included in the `sklearn.metrics` module. But for our purposes, let's just focus on the solution we get for $k = 4$, for now.

Observe that the `clusters` array we get back by calling either `.predict()` or (as in this case) `fit_predict` on a clustering estimator gives us the assigned cluster labels for each observation. Since we only used two features to do our clustering, we can easily visualize the results of the clustering in a two-dimensional scatter plot, where each point is colored based on the cluster that was assigned to it (this is what the c keyword argument to scatter means):

```
fig, ax = plt.subplots()
ax.scatter(dx_1[:, 0], dx_1[:, 1], c=clusters, edgecolor='k')
ax.scatter(*km.cluster_centers_.T, c=[0,1,2,3], edgecolor='k', linewidth=3, s=100)
ax.set_xlabel(columns[0])
ax.set_ylabel(columns[1])
g = ax.grid(None)
```

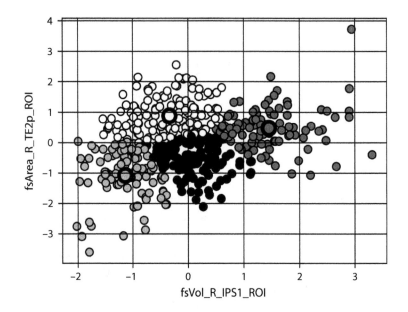

The clustering *looks* reasonable to the eye... but how far should we trust it? One way to assess clustering results is using *internal validation* methods. These methods look at the properties of the clusters and the data and quantify how well the clustering describes the observed data. For example, the Calinski-Harabasz score is quantified as the ratio between the sum of squared distances between clusters and the sum of squared distances within each of the clusters (this should remind you of the Otsu criterion used for image segmentation in Section 15.1.1). The higher this number, the more well-separated the clusters.

```
from sklearn.metrics import calinski_harabasz_score
calinski_harabasz_score(dx_1, clusters)
```

```
319.06245195929347
```

An internal validation metric such as the Calinski-Harabasz score can be used to decide which k to use, but does it really help us in understanding the clusters? Once we find an optimal k, which has the highest possible Calinski-Harabasz score, how should we think about cluster labels? Are they merely convenient descriptions of the data that facilitate the exploration of differences between subgroups of participants, or are we gaining deep insights into the true neural bases of different subtypes of autism? Another approach to validate the results of clustering is via *external validation*. In this approach, some other data is used to determine whether the results of clustering make sense. For example, if the clusters represent meaningful differences in brain structure, we might hypothesize that the clusters could also differ in some other measurement. That is, we might validate the clustering of brain properties by looking at behavioral differences between participants that end up in each of the different clusters. Ultimately, how to think about clustering results can be a thorny question and not one that we will fully answer here, but it is an important one to think about whenever one applies clustering methods to complex real-world data.

18.5   Additional Resources

As we mentioned in the text, in machine learning, we are usually more interested in quantifying the predictive capacity of a particular model. However, we acknowledge that sometimes what you want is to fit and evaluate parametric models. This is possible to do in Python using a package called StatsModels.[2] Another package that elegantly implements many of the statistical methods that behavioral scientists traditionally use (e.g., analysis of variance, or ANOVA) is Pingouin.[3]

To learn more about some of the concepts that we introduced here (e.g., Gaussian Naive Bayes and *k*-means clustering), we would particularly recommend the introductory statistical learning text *An Introduction to Statistical Learning*.[4] The book is an excellent introduction and is much more comprehensive than we could be in this chapter. Among its many merits is also the fact that it is available to download for free from the book website.

Many software packages implement machine learning methods for the analysis of neuroimaging data. One that stands out as particularly compatible with the Scikit-learn philosophy and API is the Nilearn.[5] One reason for this compatibility is that many of the contributors to the Nilearn software are also developers of Scikit-learn.

2. https://www.statsmodels.org
3. https://pingouin-stats.org/
4. https://www.statlearning.com/
5. https://nilearn.github.io

19

Overfitting

Let's pause and take stock of our progress. In the last chapter, we developed three fully operational machine-learning workflows: one for regression, one for classification, and one for clustering. Thanks to Scikit-learn, all three involved very little work. In each case, it took only a few lines of code to initialize a model, fit it to some data, and use it to generate predictions or assign samples to clusters. This all seems pretty great. Maybe we should just stop here, pat ourselves on the back for a job well done, and head home for the day satisfied with the knowledge we have become proficient machine learners.

While a brief pat on the back does seem appropriate, we probably should not be *too* pleased with ourselves. We have learned some important stuff, yes, but a little bit of knowledge can be a dangerous thing. We have not yet touched on some concepts and techniques that are critical if we want to avoid going off the rails with the skills we have just acquired. And going off the rails, as we will see, is surprisingly easy.

To illustrate just *how* easy, let's return to our brain-age prediction example. Recall that when we fitted our linear regression model to predict age from our brain features, we subsampled only a small number of features (five, to be exact). We found that, even with just five features, our model was able to capture a nontrivial fraction of the variance in age—about 20%. Intuitively, you might find yourself thinking something like this: *if we did that well with just five out of 1,440 features selected at random, imagine how well we might do with more features!* And you might be tempted to go back to the model-fitting code, replace the n_features variable with some much larger number, rerun the code, and see how well you do.

Let's go ahead and give in to that temptation. In fact, let's give in to temptation systematically. We will refit our linear regression model with random feature subsets of different sizes and see what effect that has on performance. We start by setting up the size of the feature sets. We will run this ten times in each set size, so we can average and get more stable results.

```
n_features = [5, 10, 20, 50, 100, 200, 500, 1000, 1440]
n_iterations = 10
```

We initialize a placeholder for the results and extract the age phenotype, which we will use in all of our model fits.

```
import numpy as np
results = np.zeros((len(n_features), n_iterations))
```

Looping over feature set sizes and iterations, we sample features, fit a model, and save the score into the `results` array.

```
from sklearn.linear_model import LinearRegression
model = LinearRegression()

for i, n in enumerate(n_features):
    for j in range(n_iterations):
        X = features.sample(n, axis=1)
        model.fit(X, y)
        results[i, j] = model.score(X, y)
```

Let's plot the observed R^2 as a function of the number of predictors. We will plot a dark line for the averages across iterations and use Matplotlib's `fill_between` function to add one standard deviation error bars, even though these will be so small that they are hard to see (image on the next page).

```
import matplotlib.pyplot as plt

averages = results.mean(1)

fig, ax = plt.subplots()
ax.plot(n_features, averages, linewidth=2)
stds = results.std(1)
ax.fill_between(n_features, averages - stds, averages + stds, alpha=0.2)

ax.set_xlabel("Number of brain features used")
ax.set_ylabel("Explained variance in age ($R^2$)");
```

At first glance, this might look great: performance improves rapidly with the number of features, and by the time we are sampling 1,000 features, we can predict age perfectly! But a truism in machine learning (and in life more generally) is that if something seems too good to be true, it probably is.

In this case, a bit of reflection on the fact that our model seems able to predict age with zero error should set off all kinds of alarm bells, because based on any reasonable understanding of how the world works such a thing should be impossible. Set aside the brain data for a moment and just think about the problem of measuring chronological age. Is it plausible to think that in a sample of over 1,000 people, including many young children and older adults, not a single person's age would be measured with even the slightest bit of error? Remember, *any* measurement error should reduce our linear regression model's

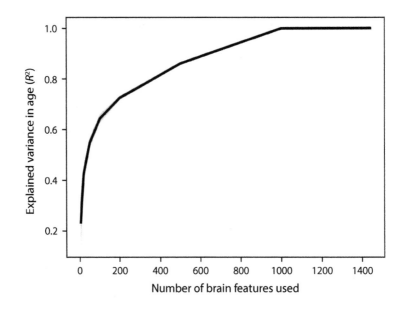

performance because measurement error is irreducible. If an Autism Brain Imaging Data Exchange II (ABIDE II) participant's birth date happened to be recorded as 1971 when it is really 1961 (oops, typo!), it is not as if our linear regression model can somehow learn to go back in time and adjust for that error.

Then think about the complexity of the brain-derived features we are using; how well (or poorly) *those* features are measured/extracted; how simple our linear regression model is; and how much opportunity there is for all kinds of data quality problems to arise. For example, is it plausible to think that all subjects' scans are of good enough quality to extract near-perfect features? If you spend a couple of minutes thinking along these lines, it should become very clear that an R^2 of 1.0 for a problem like this is just not remotely believable. There must be something very wrong with our model. And there is: our model is *overfitting* our data. Because we have a lot of features to work with (more features than samples!), our linear regression model is, in a sense, getting creative: it is finding all kinds of patterns in the data that look like they are there but are not. We will spend the rest of this section exploring this idea and unpacking its momentous implications.

19.1   Understanding Overfitting

To better understand overfitting, let's set aside our relatively complex neuroimaging data set for the moment and work with some simpler examples.

Let's start by sampling some data from a noisy function where the underlying functional form is quadratic. We will create a small function for the data generation so that we can reuse this code later on.

```
def make_xy(n, sd=0.5):
    ''' Generate x and y variables from a fixed quadratic function,
    adding noise. '''
    x = np.random.normal(size=n)
    y = (0.7 * x) ** 2 + 0.1 * x + np.random.normal(10, sd, size=n)
    return x, y
```

We fix the seed for the generation of random numbers and then produce some data and visualize them.

```
np.random.seed(10)

x, y = make_xy(30)

fig, ax = plt.subplots()
p = ax.scatter(x, y)
```

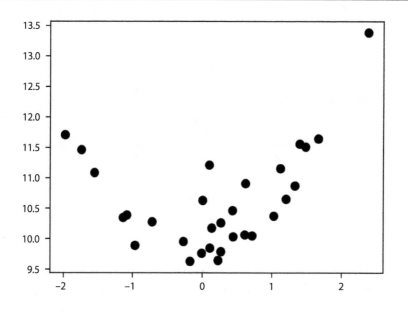

19.1.1   Fitting Data with Models of Different Flexibility

Now let's fit that data with a linear model. To visualize the results of the linear model, we will use the model that has been fit to the data to predict the y values of a range of values similar to the values of x.

```
from sklearn.metrics import mean_squared_error

est = LinearRegression()
est.fit(x[:, None], y)
```

```
x_range = np.linspace(x.min(), x.max(), 100)
reg_line = est.predict(x_range[:, None])

fig, ax = plt.subplots()
ax.scatter(x, y)
ax.plot(x_range, reg_line);
mse = mean_squared_error(y, est.predict(x[:, None]))
ax.set_title(f"Mean squared error: {mse:.2f}");
```

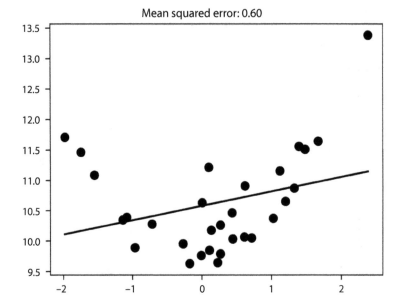

The fit looks . . . not so great. It seems pretty clear that our linear regression model is *under*fitting the data, meaning there are clear patterns in the data that the fitted model fails to describe.

What can we do about this? Well, the problem here is that our model is insufficiently flexible; our straight regression line cannot bend itself to fit the contours of the observed data. The solution is to use a more flexible estimator! A linear fit will not cut it; we need to fit *curves*.

Just to make sure we do not underfit again, let's use a much more flexible estimator, specifically a tenth-degree polynomial regression model.

This is also a good opportunity to introduce a helpful object in Scikit-learn called a Pipeline. The idea behind a Pipeline is that we can stack several transformation steps together in a sequence, and then cap them off with an Estimator of our choice. The whole pipeline will then behave like a single Estimator, i.e., we only need to call fit() and predict() once.

Using a pipeline will allow us to introduce a preprocessing step before the LinearRegression model receives our data, in which we create a bunch of polynomial features (by taking x^2, x^3, x^4, and so on—all the way up to x^{10}). We will make use of Scikit-learn's handy `PolynomialFeatures` transformer, which is implemented in the `preprocessing` module.

We will create a function that wraps the code to generate the pipeline, so that we can reuse that as well.

```python
from sklearn.preprocessing import PolynomialFeatures
from sklearn.pipeline import Pipeline

def make_pipeline(degree=1):
    """Construct a Scikit Learn Pipeline with polynomial features and
    ↪   linear regression """
    polynomial_features = PolynomialFeatures(degree=degree,
    ↪   include_bias=False)

    pipeline = Pipeline([
        ("polynomial_features", polynomial_features),
        ("linear_regression", LinearRegression())
    ])
    return pipeline
```

Now we can initialize a pipeline with `degree=10`, and fit it to our toy data:

```python
degree = 10
pipeline = make_pipeline(degree)
pipeline.fit(x[:, None], y)
reg_line = pipeline.predict(x_range[:, None])

fig, ax = plt.subplots()
ax.scatter(x, y)
ax.plot(x_range, reg_line)
mse = mean_squared_error(y, pipeline.predict(x[:, None]))
ax.set_title(f"Mean squared error: {mse:.2f}");
```

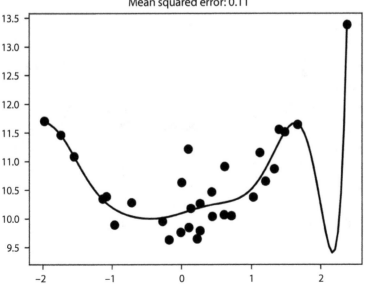

At first blush, this model seems to fit the data much better than the first model, in the sense that it reduces the mean squared error (MSE) considerably relative to the simpler linear model (our MSE went down from 0.6 to 0.11). But, much as it seemed clear that the previous model was *underfitting*, it should now be intuitively obvious that the tenth-degree polynomial model is *overfitting*. The line of best fit plotted in this example has to bend in some fairly unnatural ways to capture individual data points. While this helps reduce the error for these particular data, it is hard to imagine that the same line would still be very close to the data if we sampled from the same distribution a second or third time.

We can test this intuition by doing exactly that: we sample some more data from the same process and see how well our fitted model predicts the new scores.

```
test_x, test_y = make_xy(30)

x_range = np.linspace(test_x.min(), test_x.max(), 100)
reg_line = pipeline.predict(x_range[:, None])

fig, ax = plt.subplots()
ax.scatter(test_x, test_y)
ax.plot(x_range, reg_line)

mse = mean_squared_error(y, pipeline.predict(test_x[:, None]))
ax.set_title(f"Mean squared error: {mse:.2f}");
```

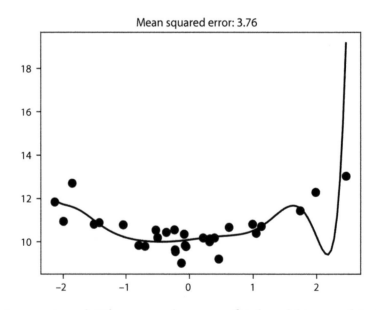

That is . . . not good. When we apply our overfitted model to new data, the MSE skyrockets. The model performs significantly worse than even our underfitted linear model did.

Exercise

Apply the linear model to new data in the same manner. Does the linear model's error also increase when applied to new data? Is this increase smaller or larger than what we observe for our tenth-order polynomial model?

Of course, since we wrote the data-generating process ourselves, and hence know the ground truth, we may as well go ahead and fit the data with the true functional form, which in this case is a polynomial with degree equal to two.

```
degree = 2
pipeline = make_pipeline(degree)
pipeline.fit(x[:, None], y)
x_range = np.linspace(x.min(), x.max(), 100)
reg_line = pipeline.predict(x_range[:, None])

fig, ax = plt.subplots()
ax.scatter(x, y)
ax.plot(x_range, reg_line)
mse = mean_squared_error(y, pipeline.predict(x[:, None]))
ax.set_title(f"Mean squared error: {mse:.2f}");
```

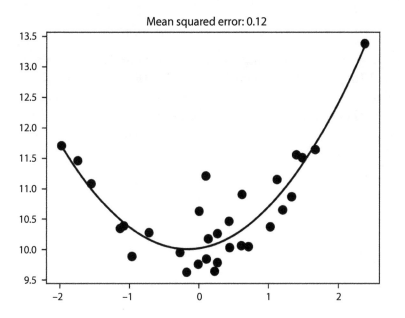

There, that looks much better. Unfortunately, in the real world, we do not usually know the ground truth. If we did, we would not have to fit a model in the first place! So we are forced to navigate between the two extremes of overfitting and underfitting in some other way. Finding this delicate balance is one of the central problems of machine learning, perhaps *the* central problem. For any given data set, a more flexible model will be able to capture more nuanced, subtle patterns in the data. The cost of flexibility, however, is that such a model is also more likely to fit patterns in the data that are only there because of noise, and hence will not generalize to new samples. Conversely, a less flexible model is only capable of capturing simple patterns in the data. This means it will avoid diving down rabbit holes full of spurious patterns, but it does so at the cost of missing out on a lot of real patterns too.

One way to think about this is that, as a data scientist, the choice you face is seldom between good models and bad ones, but rather between lazy and energetic ones (later on, we will also see that there are many different ways to be lazy or energetic). The simple linear model we started with is relatively lazy: it has only one degree of freedom to play with.

The tenth-degree polynomial, by contrast, is hyperactive and sees patterns everywhere, and if it has to go out of its way to fit a data point that is giving it trouble, it will happily do that.

Getting it right in any given situation requires you to strike a balance between these two extremes. Unfortunately, the precise point of optimality varies on a case-by-case basis. Later on, we will connect the ideas of overfitting versus underfitting (or, relatedly, flexibility versus stability) to another key concept: the *bias-variance tradeoff*. But first, let's turn our attention in the next chapter to some of the core methods machine learners use to diagnose and prevent overfitting.

19.2   Additional Resources

In case you are wondering whether overfitting happens in real research you should read the paper "I Tried a Bunch of Things: The Dangers of Unexpected Overfitting in Classification of Brain Data" by Mahan Hosseini and colleagues [Hosseini et al. 2020]. It nicely demonstrates how easy it is to fall into the trap of overfitting when doing data analysis on large complex datasets, and offers some rigorous approaches to help mitigate this risk.

20

Validation

Now that we have a basic handle on overfitting, we can start thinking about ways to mitigate this problem. In this section, we will cover methods for validating our models, i.e., obtaining better estimates of their true capabilities and identifying situations where they are behaving in undesirable ways. While model validation does not *directly* prevent overfitting, it plays an important indirect role. If we do not have a good way to recognize overfitting when it happens, we are not going to be in a great position to mitigate it.

20.1 Cross-Validation

An important insight we introduced in the last section is that a model will usually perform better when evaluated on the same data it was trained on than when evaluated on an entirely new data set. Since our models are of little use to us unless they can generalize to new data, we should probably care much more about how a model performs on new data than on data it is already seen. Basically, we want what is known as an *out-of-sample* estimate of performance.

The most straightforward way to estimate out-of-sample performance is to ensure that we always train and evaluate our model on separate, independent data sets. The performance estimate obtained from the training data set—the data set used to fit the model—will typically suffer from overfitting to some degree; an estimate on the test data set—the data set that was set aside in advance for this purpose—will not, so long as the noise in this data is independent of that in the training data set.

In practice, an easy way to construct training and test data sets with independent errors is to randomly split a data set in two. We will continue working with the Autism Brain Imaging Data Exchange II (ABIDE II) data that we started working with previously. Now, we will make use of Scikit-learn's `train_test_split` utility, found in the `model_selection` module of the library to do the work for us. The way this function works is by splitting an arbitrary number of array-like objects into training and testing subsets. For every array we pass to `train_test_split`, we get back two: a training set, and a test set. The `train_size` parameter controls the proportion of all cases assigned to the training set, with the remainder assigned to the test set. In the call shown subsequently, X and y will each contain a randomly selected half of the data set.

One part of the data get designated as training data and the other part get designated as test data.

```
from sklearn.model_selection import train_test_split
X_train, X_test, y_train, y_test = train_test_split(features, y, train_size=0.5)
```

The practice of fitting our estimator to the training data, and then evaluating its performance on the test data is called *cross-validation*, and it is ubiquitous in machine learning. Here, we will evaluate the result with both the training and the test data. The performance difference between the two will tell us how badly we are overfitting the training data. For example, we can use a linear regression model to predict age from all 1,440 structural MRI features in the ABIDE II data set. Notice that at the end we are now calculating the R^2 separately for the training sample and the testing sample:

```
from sklearn.linear_model import LinearRegression
from sklearn.metrics import r2_score

est = LinearRegression()

est.fit(X_train, y_train)

print(f"R^2 in training sample: ", est.score(X_train, y_train))
print(f"R^2 in test sample: ", est.score(X_test, y_test))
```

```
R^2 in training sample:  1.0
R^2 in test sample:  -0.9249402815055785
```

The difference here is remarkable. In the training sample, the fitted model explains 100% of the variance. In the test sample, it explains... well, none. The R^2 value is negative, implying that our model is predictively worthless! If you are used to computing R^2 by taking the square of a correlation coefficient, you might think an R^2 value below zero must be an error, but it is not. The standard definition of R^2 as the coefficient of determination is one minus the ratio of the model mean squared errors (MSE) to the variance of the data. This allows arbitrarily large negative values because the MSE can be larger than the variance if the model is a sufficiently poor fit. Intuitively, we can have an estimator that is so bad, we would have been better off just using the mean of the new data as our prediction, in which case the MSE would be identical to the variance, based on the definition of the variance, and $R^2 = 0$.

20.1.1   When Less Is More

The reason our linear regression model overfits the data is that, just as in our earlier simulated example, the model has too much flexibility: the number of features is large relative to the number of available training samples. Some neuroimaging researchers might find this statement a bit surprising, seeing as the training data set still contains over 500 subjects, and that is a pretty large sample by the standards of most MRI experiments. But the absolute number of subjects has little bearing on the propensity of a model to overfit; what matters most is the *ratio* between the model's effective degrees of freedom and the number of samples. In our case, even if we think 500 subjects is a lot (as we well might, if we are used to working with small MRI samples), we still have far more features (1,440).

If our model is overfitting because it has too many features and not enough training data, a simple way to mitigate overfitting should be to use fewer features. Let's see what happens if we randomly sample just 200 of the 1,440 features. That way we will have more samples than features.

```
X = features.sample(200, axis='columns', random_state=99)
X_train, X_test, y_train, y_test = train_test_split(X, y,
↪   train_size=0.5, random_state=99)
est.fit(X_train, y_train)
print(f"R^2 in training sample: ", est.score(X_train, y_train))
print(f"R^2 in test sample: ", est.score(X_test, y_test))
```

```
R^2 in training sample:  0.791561329658123
R^2 in test sample:  0.4576792069130101
```

That looks much better! Notice that there is still a big gap between in-sample (i.e., training) and out-of-sample (i.e., test) performance. But at least now we know that our model *is* capable of predicting age reasonably well in subjects it has not seen before.

20.1.2   K-Fold Cross-Validation

Splitting our data into training and test sets is a great way to evaluate our model's out-of-sample performance, but it comes at a cost: it *increases* the model's propensity to overfit the training data because we have halved our training sample. As a result, our model has less data to work with, which means it will be more likely to capitalize on chance and fit noise. So, at first glance, it looks like we are stuck in a catch-22: if we use all of our data to train the model, our estimate of the model's true (i.e., out-of-sample) predictive performance is likely to be heavily biased. Conversely, if we split our data into separate training and testing subsets, we get a less biased estimate of the model's true performance, but the cost is that our model will not perform as well because we are giving it less data to learn from.

Is there a way to have our cake and eat it too? In this case, as it turns out, yes! The solution is to use a form of cross-validation known as *k*-fold cross-validation. The idea here is very similar to splitting our data into training and testing halves. If we set *k* (a parameter that represents the number of *folds*, or data subsets) to two, we again end up with two discrete subsets of the data.

But now, there is an important twist: instead of using half of the data for training and the other half for testing, we are going to use both halves for training and for testing. The key is that we will take turns. First, we will use Half 1 to train, and Half 2 to test; then, we will reverse the process. Our final estimate of the model's out-of-sample performance is obtained by averaging the performance estimates we got from the two testing halves. In this way, we have managed to use every single one of our data points for both training and testing but, critically, never for both at the same time.

Of course, we do not have to set k to two; we can set it to any other value between two and the total sample size n. At the limit, if we set $k = n$, the approach is called *leave-one-out cross-validation* (because in every fold, we leave out a single data point for testing, and use the rest of the data set for training). In practice, k is most commonly set to a value in a range between three and ten. There is a long-running debate over what values of k are optimal under what conditions, and why, but we will not get into that here. (If you ever find yourself in a situation where your results change nontrivially depending on the value of k you pick, you might want to take that as a sign that your training data set is too small.)

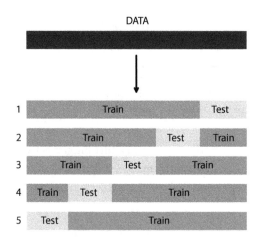

20.1.3   K-Fold the Explicit Way

To illustrate how k-fold cross-validation works, let's implement it ourselves. First, we create k different subsets of the original data set. Then, we loop over the k subsets and, in each case, use the current subset to test the model trained on the remaining $k - 1$ subsets. Finally, we average over the performance estimates obtained from all k folds to obtain our overall out-of-sample performance estimate. If you are not interested in wading through the code, you can skip to the next subsection, where we replace most of this with a single line.

First, let's set a value for k

```
k = 5
```

We initialize placeholders for the results using `numpy.zeros()` and initialize the `LinearRegression` object we will use for fitting.

```
import numpy as np
train_r2 = np.zeros(5)
test_r2 = np.zeros(5)
est = LinearRegression()
```

Python comes with a library for the generation of randomness (getting a computer to behave randomly is trickier than you think). Here, we will use the shuffle function from this library, which takes a list and reorganizes it in random order. In this case, we will organize the numbers from zero thru the number of rows in X.

```
from random import shuffle
indices = list(range(len(X)))
shuffle(indices)
```

We loop over the folds. In each fold, we select every k-th index to belong to the training set. Next, we designate all the indices not chosen to be test-set samples as training indices (we use a list comprehension, see chapter 5 if you need a reminder). The linear regression is fit to the training data. In each iteration, we score the model with both the training and the testing data. When we are done, we have five estimates of R^2 based on five fits of the model.

```
for i in range(k):
    test = indices[i::k]
    train = [jj for jj in indices if jj not in test]
    X_train = X.iloc[train]
    X_test = X.iloc[test]
    y_train = y.iloc[train]
    y_test = y.iloc[test]

    est.fit(X_train, y_train)

    train_r2[i] = est.score(X_train, y_train)
    test_r2[i] = est.score(X_test, y_test)

print("Fold scores in training: ", train_r2)
print("Mean training R^2 over folds: ", np.mean(train_r2))

print("Fold scores in test: ", test_r2)
print("Mean training R^2 over folds: ", np.mean(test_r2))
```

```
Fold scores in training:  [0.75885782 0.76459993 0.74761466 0.74439135 0.75183613]
Mean training R^2 over folds:  0.7534599775648393
Fold scores in test:  [0.5398583  0.53337957 0.57794919 0.59105246 0.57138778]
Mean training R^2 over folds:  0.5627254590096887
```

Notice that the results of the five-fold cross-validation differ from those we got when we used 50% of the data to train the model and 50% of them to test the model. Our in-sample performance is lower now (0.79 versus 0.75), but our out-of-sample performance, which is what we care about, is higher (0.46 versus 0.55).

Exercise

Why do you think out-of-sample accuracy is higher in this setting? What is the downside of having higher average out-of-sample accuracy in this case?

20.1.4   K-Fold the Easy Way

The *k*-fold cross-validation is an extremely common validation strategy, so any machine learning package worth its salt should provide us with some friendly tools we can use to avoid having to reimplement the basic procedure over and over. In Scikit-learn, the `cross_validation` module contains several useful utilities. We have already seen `train_test_split`, which we could use to save us some time. But if all we want to do is get cross-validated scores for an estimator, it is even faster to use the `cross_val_score` function. The function takes an estimator, the inputs to the `fit` method, and an optional specification of the cross-validation procedure. Integers are interpreted as the number of folds to use in a *k*-fold partitioning (but there are other possible inputs to the `cv` keyword argument, which we will not discuss here).

```
from sklearn.model_selection import cross_val_score
r2_cv = cross_val_score(est, X, y, cv=k)

print("Individual fold scores:", r2_cv)
print(f"\nMean cross-validated R^2: ", np.mean(r2_cv))
```

```
Individual fold scores: [0.60389708 0.53795077 0.56791833 0.56837218 0.55742804]

Mean cross-validated R^2:   0.5671132813920163
```

That is it! We were able to replace nearly all of our code from the previous example with one function call. If you find this a little *too* magical, Scikit-learn also has a bunch of other utilities that offer an intermediate level of abstraction. For example, the `sklearn.model_selection.KFold` class will generate the folds for you, but will return the training and test indices for you to loop over, rather than automatically cross-validating your estimator.

20.1.5   Cross-Validation Is Not a Panacea

There is a reason cross-validation is so popular: it provides an easy way to estimate the out-of-sample performance of just about any estimator, with minimal bias or loss of efficiency. But it is not magic. One problem it does nothing to solve is that of interpretation. It is tempting to conclude from our healthy R^2 above that we have shown that there is some important causal relationship between brain structure and chronological age. Unfortunately, the fact that our model seems to predict age pretty well does not mean we know *why* it is predicting age pretty well. We do not even know that its predictive power derives from anything to do with the brain per se. It is possible that our brain-derived features just happen to be correlated with other non-brain features we did not measure, and that it is the omitted variables that are doing the causal work for us. This might not bother us if we *only* cared about making good predictions. But it could be a very serious problem if we are using machine learning as an instrument to advance scientific understanding. If it turns out that the brain features in our data set are highly correlated with some other variable(s), that might radically change our interpretation of what our model is doing.

As it happens, the brain variables in our data set *are* correlated with at least one variable that is very strongly associated with age: research site. The ABIDE II data set we are working with contains scans from seventeen sites. Let's look at how age varies by site (we will use the Seaborn library for the visualization; image on the next page).

```
import seaborn as sns
import matplotlib.pyplot as plt
g = sns.catplot(x='site', y='age', data=abide_data, kind='strip', height=4, aspect=4)
g = g.set_xticklabels(rotation=30, horizontalalignment='right')
```

In the previous plot, each strip is a different site, and each point is a single subject. As we can clearly see, sites do indeed vary wildly in their age distributions, for example, the NYU site scanned only very young children (< 10), whereas the BNI site scanned mostly adults over age 40.

The fact that site differences are so big is worrisome in one sense, as it raises the possibility that at least some of the predictive power of our model might derive from spurious associations between the research site and our brain features, i.e., that if we were to adjust age to account for between-site differences, the brain features might no longer do quite as much work for us in predicting age.

Exercise

There are various ways we can go about adjusting age prediction for research site. A simple but very conservative one is to residualize age on research site; i.e., is, we remove all the variance in the age that can be explained by between-site differences. To save time, we have included an already-residualized version of the age variable in our data set; it is in the `age_resid` column. Repeat our analysis as shown previously with `'age_resid'` instead of `'age'`. What does the result mean?

20.1.6   Is It All for Naught?

Does this mean we are fooling ourselves, and should just give up now? No! As noted previously, this particular analysis is *extremely* conservative. Because the ABIDE II sites differ so dramatically in age distributions, by controlling for research sites, we are wiping out most of the variability in age. This would make it much harder to predict age from the brain features even if it was the site effects that were completely spurious. The root problem is that we cannot tell, just from a set of correlations, what the correct causal model is.

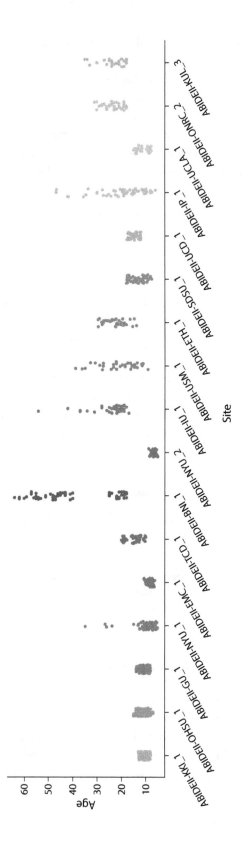

The reality is that causal interpretation of observational data is hard, and cross-validation does not make it any easier. The question of whether or not we need to adjust for variables like site effects before examining the brain-age relationship is not trivial (nor is deciding *how* to adjust for them). For now, we will ignore the effects of the research site, or any other potential confounders, and just try to predict age directly from the brain variables. Just be aware that if we were planning to share these results with the rest of the world (e.g., in a paper reporting what we have found), we would probably need to think much more carefully about our causal assumptions, and maybe do some extra work to convince ourselves that there genuinely is a good deal of information about chronological age being carried by structural differences in people's brains. We will come back to related issues toward the end of the book.

20.2   Learning and Validation Curves

So far, we have focused on cross-validating model performance at a single point in the data set and parameter space. That is to say, we are asking the following: How well would our fitted model do if we took it exactly as-is, changing none of its parameters, and applied it to a new data set sampled from the same population?

This is an important thing to know, but we can do better. If we want to get deeper insights into a model's behavior, it is helpful to observe the way it responds when we systematically vary the size and nature of the data, the number and type of features, and the model hyperparameters.

In Scikit-learn, we can use the `learning_curve` and `validation_curve` utilities to flexibly and compactly wrap most of the cross-validation functionality we have already encountered. The idea behind the *learning-curve* is to graphically display a model's predictive performance as a function of the model's experience. That is, to visualize how well it learns as the available data grows. The *validation_curve* is directly analogous, except instead of varying the size of the data, we systematically vary one of the estimator's parameters.

Let's look at `learning_curve` in more detail. Here is an example:

```
from sklearn.model_selection import learning_curve
```

We will use 100 random features as predictors.

```
X_subset = X.sample(100, axis=1, random_state=100)
```

We will assign a few different sizes of the samples we want to plot model performance for and use a `LinearRegressor`. Note that we cannot go above 800, because we only have ~1,000 cases, and we are using 80% of the sample for training.

```
train_sizes = [100, 200, 400, 800]
est = LinearRegression()
```

Next, we obtain the learning curve. as with the `cross_val_score` utility, the cross-validated application of our estimator is done implicitly for us.

```
results = learning_curve(est, X_subset, y, train_sizes=train_sizes,
                         cv=5, shuffle=True)
sizes, train_scores, test_scores = results
```

The `train_scores` and `test_scores` arrays contain the performance scores for training and testing data, respectively. The rows reflect training sizes (i.e., in our case, the first row gives performance $n = 100$, the second for $n = 200$) and the columns contain the scores from the k folds of the k-fold cross-validation procedure.

```
test_scores.round(2)
```

```
array([[-3.7030e+01, -1.5611e+02, -2.4600e+00, -1.2440e+01, -3.8460e+01],
       [ 2.4000e-01,  2.2000e-01,  1.0000e-01,  3.0000e-01,  2.7000e-01],
       [ 4.9000e-01,  4.4000e-01,  5.5000e-01,  5.3000e-01,  4.7000e-01],
       [ 6.0000e-01,  5.7000e-01,  5.8000e-01,  5.5000e-01,  5.8000e-01]])
```

This information is fairly hard to appreciate in tabular form, so let's average performance over the columns (i.e., the k folds) and plot it as a function of the sample size (image on the next page).

```
train_mean = train_scores.mean(1)
test_mean = test_scores.mean(1)

fig, ax = plt.subplots()
ax.plot(sizes, train_mean, 'o-', label='Training')
ax.plot(sizes, test_mean, 'o-', label='Test');
ax.grid(True, axis='y')
ax.legend(fontsize=14)
ax.set_ylim(0, 1)
ax.set_xlabel('Sample size (n)')
ax.set_ylabel('$R^2$');
```

We learn several things from this plot. First, our linear regression estimator overfits with small samples: R^2 is near one in the training sample, but is essentially zero in the test sample. Second, R^2 in the test sample improves monotonically as the training data set gets larger. This is a general rule: on average, a model's out-of-sample performance should only get better as it is trained on more data. Third, note that the training and test curves never converge, even with the largest sample we have available. This suggests that we are

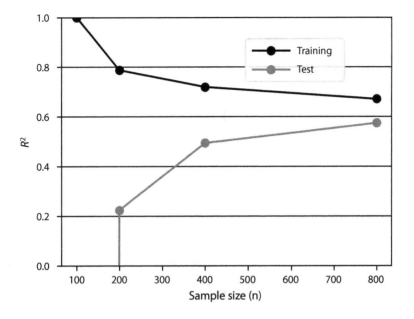

still overfitting, and we might want to consider either acquiring more data (if possible) or reducing the model's flexibility (e.g., by using fewer features, or, as we will see in later sections, by using a different model less prone to overfitting).

20.2.1   Training versus Test versus Validation: A Note on Nomenclature

Before moving on, we have a quick note on nomenclature: you will sometimes hear people talk about validation sets rather than (or in addition to) training and test sets. What this means depends on the context. In many cases, the terms test set and validation set are used interchangeably to refer to a new data set independently of the one the model is trained on, but sometimes the distinction between validation and test is important. In particular, it is common to set aside a true hold-out data set in advance of any validation efforts. In such cases, the test data set is the one we use to obtain our final estimate of performance. The validation data set, by contrast, is technically part of the training data (in that we allow ourselves to use it to train the final model), but it is being used for validation. When we perform a cross-validation procedure on a training set, we call the hold-out folds the validation sets.

This kind of three-way split of our data is an excellent way to operate, as it ensures that we are able to detect overfitting that occurs not only during model estimation but also in the model selection process. For example, if we cross-validate 100 models and then choose the one with the best cross-validated performance, we are still going to overfit to some degree, and performance in the test data set will reveal this. Next, let's consider how we can go one further step toward systematically and automatically choosing the right model for our data.

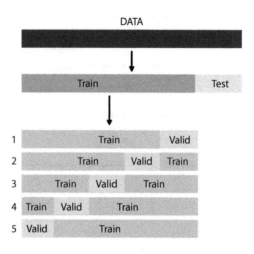

Exercise

We have looked at the relationship between training data set size and model performance. Here, you will add another dimension to the picture, and try to get a qualitative sense of how our training and validation curves vary as a function of model complexity. In this case, we will operationalize this as the number of predictors our linear regression estimator gets to use.

Use the ABIDE II data to predict age given three different sizes of feature sets: five features, thirty features, and 100 features, which you select at random. Once again, evaluate performance for sample sizes ranging from 100 to 800.

Does our ability to predict age vary depending on how many features we use? Do the characteristics of overfitting change with the sample size?

We will return to a discussion of model complexity in the final chapter of this part of the book (chapter 22) when we consider the complexity of deep learning models with many thousands of parameters that can be adjusted.

20.3   Additional Resources

Cross-validation is a very powerful tool, but it does not address all of the challenges of understanding your data. For example, as demonstrated in a paper by Gaël Varoquaux [Varoquaux 2017], even when you do use cross-validation, the errors in your estimate of accuracy can be rather large, particularly if the sample size is not large enough. The paper is worth reading and pondering when designing studies that use machine learning methods and considering the sample size one might need.

21

Model Selection

Machine learning practitioners often emphasize the importance of having very large training data sets. It is all well and good to say that if we want to make good predictions, we should collect an enormous amount of data; the trouble is that sometimes that is not feasible. For example, if we are neuroimaging researchers acquiring structural brain scans from people with an autism diagnosis, we cannot magically make tens of thousands of scans appear out of thin air. We may only be able to pool together one or two thousand data points, even with a monumental multisite effort like the Autism Brain Imaging Data Exchange (ABIDE) initiative.

If we cannot get a lot more data, are there other steps we can take to improve our predictions? We have already seen how cross-validation methods can indirectly help us make better predictions by minimizing overfitting and obtaining better estimates of our models' true generalization abilities. In this section, we will explore another general strategy for improving model performance: selecting better models.

21.1   Bias and Variance

We have talked about the tradeoff between overfitting and underfitting, and the related tradeoff between flexibility and stability in estimation. Now we will introduce a third intimately related pair of concepts: bias and variance. Understanding the difference between bias and variance, and why we can often trade one for the other, is central to developing good intuitions about when and how to select good estimators.

It is probably easiest to understand the concepts of bias and variance through visual illustration. A classic representation of these concepts appears in the image on the following page.

Comparing the top and bottom rows gives us a sense of what we mean by *variance*: it is the degree of scatter of our observations around their central tendency. In the context of statistical estimators, we say that an estimator has high variance if the parameter values it produces tend to vary widely over the space of data sets to which it could be applied. Conversely, a low-variance estimator tends to produce similar parameter estimates when applied to different data sets.

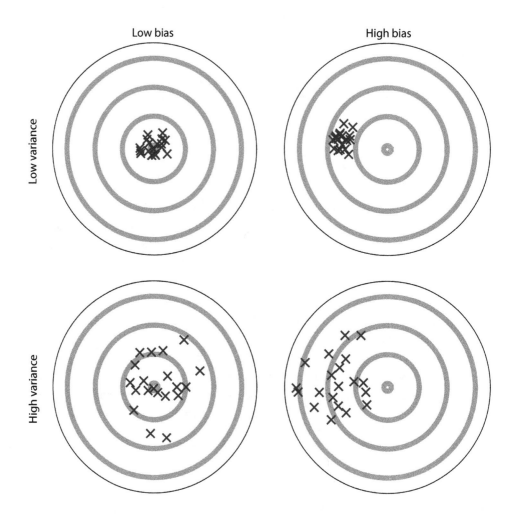

Bias, by contrast, refers to a statistical estimator's tendency to produce parameter estimates that cluster in a specific part of the parameter space. A high-bias estimator produces parameter estimates that are systematically shifted away from the correct estimate, even as the size of the training data set grows asymptotically. A low-bias estimator, conversely, will converge on the true parameter estimate as the sample size increases.

21.1.1   The Bias-Variance Tradeoff

One interesting property of bias and variance is that they tend to trade off. That is, one can often decrease the variance of an estimator by deliberately increasing its bias or vice versa. To see this intuitively, imagine the extreme case of an estimator that always generates an identical prediction for all observations. Such an estimator would have low variance (actually, *no* variance) and extremely high bias. Conversely, an estimator that produces extremely variable estimates must be one capable of freely exploring various parts of parameter space and thus shows a weaker preference for, or bias toward, certain parts of the space.

This idea has a deep connection to the overfitting/underfitting and flexibility/stability distinctions we drew earlier. All three distinctions ultimately reflect the same core idea. A *high-variance* (low-bias) estimator is a *flexible* one that can identify a wide variety of patterns in the data; this also makes it more likely to *overfit*. Conversely, a high-bias (low-variance) estimator is a *stable* one that tends to produce similar parameter estimates no matter what data it is presented with; such an estimator will frequently *underfit* because it lacks the flexibility needed to capture complex patterns in the data.

Neuroimaging researchers coming from a traditional scientific background often think of bias as a dirty word, i.e., as something to be minimized at all costs. From a machine learning standpoint, however, there is nothing intrinsically bad about bias. The bias of a particular model is nothing more than its systematic tendency to produce parameter estimates that cluster within a particular part of the space. Not only is this not a bad thing, but it is often essential to producing good predictions. The bias of a given estimator— its tendency to give certain kinds of solutions—is often referred to as its *inductive bias*. Then the key question is: *To what degree does the inductive bias of a given estimator match the processes that give rise to our data?*

In cases where there is a match between an estimator and the data it is applied to, a biased estimator will tend to perform better than a less biased one. Conversely, if an estimator has the *wrong* bias, it will tend to perform more poorly than an estimator with less bias.

This is a fairly abstract idea, so let's make it concrete. Suppose you park your car at the airport parking lot when you head off for a weekend vacation. Consider two possible strategies you could use to find your car on your return. First, you might rely on your memory of where you left the car; perhaps you think you left it somewhere near that light pole over there in the B area of the lot. Second, you could just randomly roam around the parking lot looking for a car that looks like yours.

Which of these is the better car-finding strategy? Well, it depends! If you have a good memory, the high-bias strategy, i.e., only searching within a particular part of the lot, will lead you to your car more quickly than a random walk around the entire lot. But if your memory is poor, the biased strategy is likely to be counterproductive, because your car will not be where you remember leaving it, no matter how much time you spend walking around that specific area. In the latter case, you would have been better off randomly walking around until you found your car.

21.2 Regularization

Our car parking example is a bit contrived, but it illustrates a core principle at the heart of good model selection: *if we have useful prior knowledge about the domain, we should pick an estimator that builds that knowledge into its assumptions.* For example, if we have a set of 1,000 potential features we could use in our model, but have good reason to believe that only a small fraction of those features make a meaningful contribution to the outcome variable, it seems like a good idea to use an estimator that can somehow *select* a subset of features. Using an estimator that acts as if all of our features are potentially important (as linear regression does) is likely to lead to overfitting.

The principle of introducing background knowledge into our model to improve predictions and prevent overfitting is known in machine learning and statistics as *regularization*. When we say we are using a regularized estimator, what we are typically saying is that we have taken an existing estimator (e.g., an ordinary least-squares (OLS) estimator) and constrained its behavior with additional information, so that it is *biased* toward certain parts of the parameter space.

21.2.1 Penalized Regression

To researchers who learned statistics in biomedical or social science training programs, it may be tempting to think of least-squares estimation as a kind of automatic default, with all other procedures requiring some special justification to use. But there is nothing special about OLS. Well, it is *kind of* special in that it always minimizes the squared error in our training sample. But as we have already seen, we usually care about minimizing test error, not training error (plus, sometimes we do not want to quantify error in terms of a least-squares criterion). So if we can get our hands on an estimator that is biased relative to OLS, but tends to reduce error in the *test* data set, we probably still want to use it, even if it does not reduce error in the *training* data quite as well as OLS.

It turns out that there are many such estimators. One of the most widely used class of regularized estimators for regression problems is *penalized regression*. These methods generally start with garden variety OLS and then add a little twist in the form of a *penalty parameter*.

Consider the cost function we seek to minimize in OLS:

$$Cost = RSS = \sum_i^N (y_i - \sum_j^P \beta_j x_{ij})^2$$

Here, RSS is the residual sum of squares (which is the same as the sum of squared errors), N is the number of samples, and P is the number of features. Our goal is to obtain the set of β coefficients that minimize the RSS.

Now consider a slight variation:

$$Cost = \sum_i^N (y_i - \sum_j^P \beta_j x_{ij})^2 + \alpha \sum_j^P |\beta_j|$$

The only difference here is that, in addition to the RSS, our cost function includes a second term, $\alpha \sum_j^P |\beta_j|$, which is the sum of absolute values of the β coefficients weighted by the penalty parameter, α. The addition of the penalty parameter is the reason we refer to this variation on linear regression as penalized regression. The addition of the penalty may seem like a small change, but it has major consequences, which we will explore shortly.

The previous form of penalized regression is known as *lasso regression*, and it is very widely used. There is also another very common form of penalized regression known as *ridge regression.* It looks a lot like lasso regression, except instead of computing the penalty as the sum of absolute coefficients, we take the sum of their squares:

$$Cost = \sum_i^N (y_i - \sum_j^P \beta_j x_{ij})^2 + \alpha \sum_j^P \beta_j^2$$

The difference between lasso and ridge regression may seem even smaller, but it again has important implications. Let's explore the behavior of each of these methods.

LASSO REGRESSION

Let's take the lasso first. Let's think about what the addition of the penalty term to the standard RSS criterion does, conceptually. By making the overall cost depend on a (weighted) sum of absolute coefficients, we are saying that there is a certain cost to having large coefficients in our model. If the net negative contribution of a given feature to the RSS is smaller than its net positive contribution to the penalty, then it will be beneficial to shrink that feature's coefficient to the point where the two terms perfectly cancel each other out. Thus, penalized regression *biases* the estimated regression coefficients by shrinking at least some of them.

For reasons outside the scope of our discussion here, the lasso does not just shrink coefficients; it shrinks them to zero. That is, as we increase the penalty, more features drop out of our model, leaving us with a simpler prediction equation involving fewer features. For this reason, the lasso is often described as implicitly including a *feature selection* step.

To see how this works in practice, let's return to our ABIDE II data and visualize what happens to our regression coefficients estimated with lasso regression as we increase the penalty parameter. Here, we will plot the *coefficient paths*: how each coefficient changes as α changes (judiciously picking a reasonable range of α values). We have implemented a `plot_coef_path` function that does this. Note also that we standardize our features to mean-zero and unit-variance because lasso and ridge are sensitive to scale.

```
from ndslib.viz import plot_coef_path
from ndslib.data import load_data
abide_data = load_data("abide2")
features = abide_data.filter(like='fs')
phenotypes = abide_data.iloc[:, :6]

from sklearn.linear_model import Lasso
from sklearn.preprocessing import scale

number_of_features = 200

X = scale(features.sample(number_of_features, axis=1, random_state=99))
y = phenotypes['age']

import numpy as np
alpha = np.logspace(-3, 1, 100)
```

```
ax = plot_coef_path(Lasso, X, y, alpha, max_iter=2000)
ax.set_title("Coefficient paths for lasso regression")
ax.set_xlim(1e-3, 10);
```

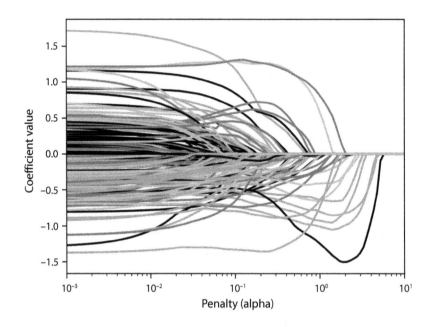

In the figure produced by the `plot_coef_path` helper function, each line repre-sents a different feature. We are using 200 randomly selected features (from the full set of 1,440) as predictors, so there are 200 lines. The x-axis displays the penalty parameter used for lasso estimation; the y-axis displays the resulting value of the coefficients. Notice how, at a certain point (around `alpha = 0.01`), coefficients start to disappear. They are not just small; they are *gone* (i.e., they have been shrunk to zero). This built-in feature selection is a very useful property of the lasso. It allows us to achieve an arbitrarily sparse solution simply by increasing the penalty.

Exercises

1. Change the `number_of_features` variable to be much larger (e.g., 400) or much smaller (e.g., 100). How are the coefficient paths affected by this change? Why do you think these changes happen?
2. What happens if you use the site-residualized ages instead of the age variable?

Of course, as you know by now, there's no free lunch in machine learning. There must be some price we pay for producing more interpretable solutions with fewer coef-ficients, right? There is. We will look at it in a moment; but first, let's talk about ridge regression.

RIDGE REGRESSION

Ridge regression, as noted previously, is mathematically very similar to lasso regression. But it behaves very differently. Whereas the lasso produces sparse solutions, ridge regression, like OLS, always produces dense ones. Ridge *does* still shrink the coefficients (i.e., their absolute values get smaller as the penalty increases). It also pushes them toward a normal distribution. The latter property is the reason a well-tuned ridge regression model usually outperforms OLS: you can think of ridge regularization as basically a way of saying "it may look like a few of the OLS coefficients are way bigger than others, but that is probably a sign we are fitting noise." In the real world, outcomes usually reflect the contributions of a lot of different factors. So let's squash all of the extreme values toward zero a bit so that all of our coefficients are relatively small and bunched together.

```
from sklearn.linear_model import Ridge

# Coefficient paths for ridge regression, predicting age from 30 features
alpha = np.logspace(-5, 5, 100)
ax = plot_coef_path(Ridge, X, y, alpha)

ax.set_title("Coefficient paths for ridge regression")
ax.set_xlim(1e-5, 1);
```

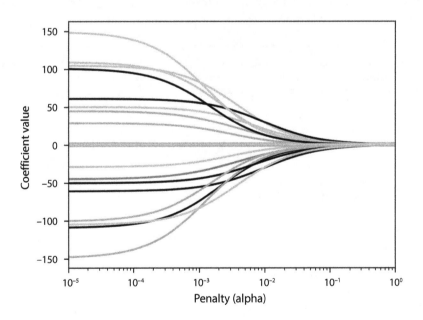

It may look to the eye like the coefficients are shrinking all the way to zero as the penalty increases, but we promise you they are not. They may get very small, but unlike lasso regression, ridge coefficients will never shrink to nothing.

EFFECTS OF REGULARIZATION ON PREDICTIVE PERFORMANCE

We have looked at two different flavors of penalized regression, and established that they behave quite differently. So, which one should we use? Which is the *better* regularization approach? Well, as with everything else, *it depends*. If the process(es) that generated our data are sparse (or at least, have a good sparse approximation), meaning, a few of our features make very large contributions and the rest do not matter much, then lasso regression will tend to perform better. If the data-generating process is dense, i.e., lots of factors make small contributions to the outcomes, then ridge regression will tend to perform better.

In practice, we rarely know the ground truth when working with empirical data. So we are forced to rely on a mix of experience, intuition, and validation analyses.

Let's do some of the latter here. To try to figure out what the optimal lasso and ridge penalties are for our particular problem, we will make use of Scikit-learn's `validation_curve` utility, which allows us to quickly generate training and testing performance estimates for an estimator as a function of one of its parameters. As alluded to in chapter 20, the `validation_curve` is very similar to the `learning_curve` we have seen before, except that instead of systematically varying the data set size, we systematically vary one of the estimator's parameters (in this case, `alpha`). Let's generate a validation curve for lasso regression, still using 200 random features to predict age. We will add OLS regression R^2 for comparison and use our `plot_train_test` helper function to visualize it (image appears on the next page).

```
from sklearn.model_selection import validation_curve, cross_val_score
from sklearn.linear_model import LinearRegression
from ndslib.viz import plot_train_test

x_range = np.logspace(-3, 1, 30)
train_scores, test_scores = validation_curve(
    Lasso(max_iter=5000), X, y, param_name='alpha',
    param_range=x_range, cv=5, scoring='r2')

ols_r2 = cross_val_score(
    LinearRegression(), X, y, scoring='r2', cv=5).mean()

ax = plot_train_test(
    x_range, train_scores, test_scores, 'Lasso',
    hlines={'OLS (test)': ols_r2})

ax.set_title("Age prediction performance")
ax.set_xlim(1e-3, 10)
```

Notice that, in the training data set, R^2 decreases monotonically as alpha increases. This is necessarily true. Why do you think that is? (Hint: think of how R^2 is defined, and its relationship to the least-squares component of the lasso cost function.)

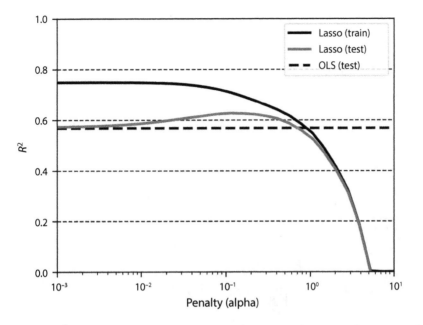

In contrast, R^2 in the test data set is highest within a specific range of `alpha` values. This optimum will vary across data sets, so you will almost always need to tune the alpha parameter (e.g., using a validation curve like the one shown in the previous example) to get the best performance out of your model. This extra degree of complexity in the model selection process is one of the reasons you might *not* want to always opt for a penalized version of OLS, even if you think it can improve your predictions. If you muck things up and pick the wrong penalty, your predictions could be way worse than OLS!

While we have only looked at one data set, these are fairly typical results; in many application domains, penalized regression models tend to outperform OLS when using small to moderate-sized samples. When samples get very large relative to the number of features, the performance difference usually disappears. Outside of domains with extremely structured and high-dimensional data, this latter principle tends to be true more globally, i.e., more data usually beats better algorithms.

Exercise

Use a similar approach to compare ridge regression with OLS.

21.3 Beyond Linear Regression

We have focused our attention here on lasso and ridge regression because they are regularized versions of the OLS estimator most social scientists are already familiar with. But just as there is nothing special about OLS, there is also nothing special about (penalized) regression in general. There are many different estimators we could use to generate

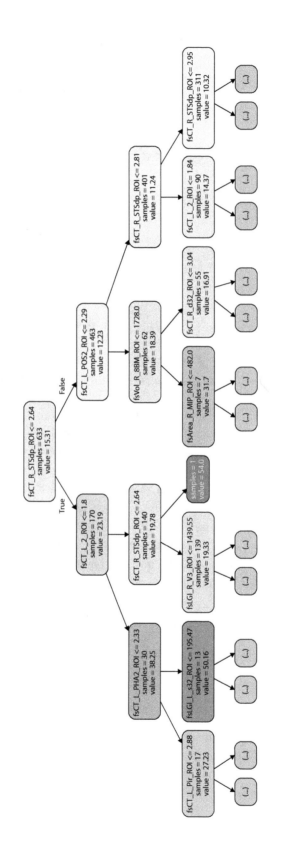

predictions. Each has its own inductive biases and will perform better on some data sets and worse on others. The science (or perhaps art) of machine learning lies in understanding the available estimators, data, and validation methods well enough to know how to tailor optimal (or at least good) machine learning workflows to the specific problem at hand. Let's forge ahead with one more widely used class of algorithms that differs quite a bit from the regression algorithms you have seen here so far.

21.3.1 Random Forests

To illustrate how easy it is to try out completely different types of estimators in Scikit-learn, and also draw attention to the fact that estimators based on linear regression represent only a small part of the space, let's repeat some of our earlier regression analyses using a very different estimator: the RandomForestRegressor. Random forests are essentially collections (or *ensembles*) of decision trees. A decision tree is a structure for generating classification (classification trees) or regression (regression trees) predictions. The figure on the previous page shows an example of what a regression tree might look like if used in our ABIDE II data set to predict age from structural brain features.

At each node, we evaluate a particular conditional. For example, the first question we ask is this: Is the subject's observed value on the 'fsCT_R_STSdp_ROI' variable less than or equal to 2.64, or is it greater? We follow the appropriate branch to the next node and answer the question we find there. We repeat this process until we reach a terminal node (or leaf), whereupon the tree produces the predicted value for that observation.

Random forests extend this idea by *bagging* multiple decision trees and aggregating their outputs. Decision trees are very flexible and have the propensity to overfit. By averaging over a lot of different trees (e.g., with each one generated by resampling the data), we hopefully stabilize our predictions and reduce overfitting. Random forests are popular because they are extremely powerful, their constituent trees are highly interpretable, and they tend to perform well when we have a lot of data. The downsides are that being very flexible, they have a tendency to overfit, and careful tuning may be required to achieve good performance. They can also be quite slow, as performance tends to improve with the number of decision trees. In the subsequent example we will reduce tree complexity and mitigate overfitting by performing an explicit feature selection step before fitting the model. This is done using the SelectKBest object and an f_regression criterion, which selects the k features that have the strongest correlation with the target (based on an F-test, hence the name). Ideally, we would want to use a larger number of trees, but this would be very slow, so we will stick with a small number (ten) for demonstration purposes (but please experiment with changing these numbers).

```
from sklearn.ensemble import RandomForestRegressor
from sklearn.feature_selection import SelectKBest, f_regression

from ndslib.viz import plot_learning_curves

number_of_features = 50
```

```
alpha = 10
number_of_trees_in_forest = 10

selector = SelectKBest(f_regression, k=number_of_features)
X = selector.fit_transform(features, y)

estimators = [Ridge(alpha), RandomForestRegressor(number_of_trees_in_forest)]
labels = ["Ridge regression", "Random forests"]

plot_learning_curves(estimators, [X, X], y, [100, 200, 500, 900], labels)
```

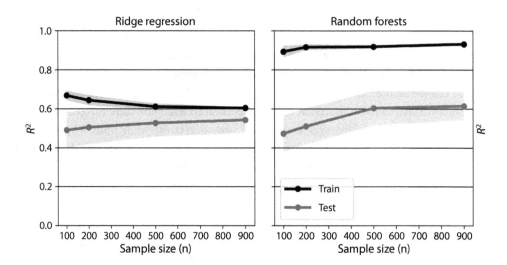

In this case, the random forest estimator slightly outperforms ridge regression, particularly as the sample size gets larger. But it is worth reiterating our earlier point: this is not some kind of law, or even a particularly good heuristic. If you play with various parameters (e.g., `number_of_features`, `alpha`), you can easily find regimes under which ridge regression performs better.

INTERPRETING RANDOM FORESTS

Unlike regression-based methods, decision trees (and hence random forests) do not have linear coefficients. Instead, we can obtain *feature importances*, which tell us how important each feature is to the overall prediction (more important features are better at reducing uncertainty, and occur closer to the root of the tree). Let's randomly sample thirty features and work with those. After fitting, we will print out the names of the ten most important features.

```
rf = RandomForestRegressor(100)
X = features.sample(30, axis=1)
rf.fit(X, y)
import pandas as pd
print(pd.Series(
    rf.feature_importances_, index=X.columns).sort_values(ascending=False).head(10))
```

```
fsCT_R_RI_ROI          0.245458
fsVol_L_a10p_ROI       0.071267
fsVol_L_EC_ROI         0.068130
fsCT_R_v23ab_ROI       0.053689
fsVol_L_33pr_ROI       0.046933
fsVol_R_IP0_ROI        0.046815
fsCT_L_PoI2_ROI        0.042008
fsVol_L_v23ab_ROI      0.038718
fsVol_L_3a_ROI         0.032332
fsCT_L_47s_ROI         0.031544
dtype: float64
```

Be aware that there are some important subtleties related to the interpretation of feature importances, not least of which is that, as with any model, feature importances are configural: a feature may seem important when modeled alongside one set of other features, but unimportant when included with a different set of features.

VISUALIZING TREES

We can also plot the individual decision trees. Note that different trees can often exhibit very different structures despite performing equally well predictively, an observation that should lead us to exercise caution when interpreting decision trees, even if they *seem* straightforward. In this particular case, as we saw previously, a handful of features dominate the list of feature importances, so we might expect at least the first few nodes to look relatively similar. Let's take a look at the first two trees in the forest. Indeed, the feature that had the highest feature importance (fsCT_R_RI_ROI) appears in one of the top levels of both trees (images appear on the following pages).

```
from ndslib.viz import plot_graphviz_tree
plot_graphviz_tree(rf.estimators_[0], X.columns)
plot_graphviz_tree(rf.estimators_[1], X.columns)
```

21.4 Additional Resources

To learn more about the properties of the lasso and ridge algorithms, we would recommend reading an excellent tutorial by Ryan Tibshirani that Larry Wasserman amended and posted on his website.[1] If you want to think a bit more about the implications of the α parameter, particularly in ridge regression, you can read a paper that one of us wrote on the topic [Rokem and Kay 2020].

---

1. https://www.stat.cmu.edu/~larry/=sml/sparsity.pdf

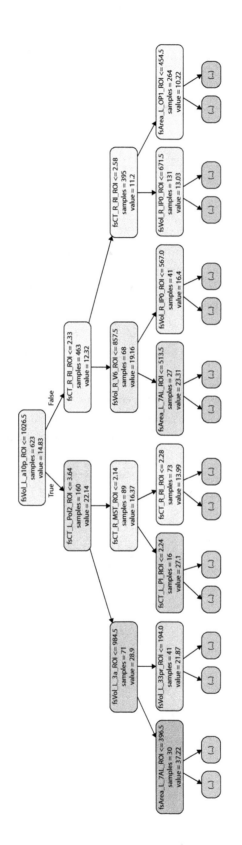

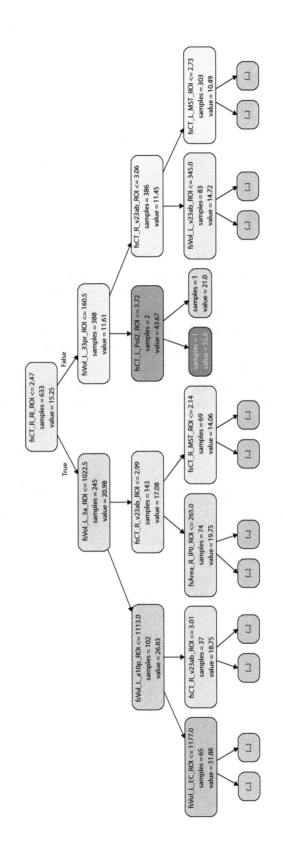

22

Deep Learning

While we promised that this part of the book would not be a catalog of different machine learning methods, no book introducing machine learning—and particularly applied to a topic so closely bound with image processing—would be complete without at least a brief mention of deep learning. Deep learning is a relatively new branch of machine learning that encompasses a set of techniques that combine advances in artificial neural networks with clever optimization methods and large amounts of data to do remarkably well at many different machine learning tasks. We would hesitate to call these methods artificial intelligence (AI), but these algorithms are prone to doing things that until recently we would have thought only humans can do, like recognize objects in a photograph, generate (at least somewhat) cogently written prose, or play Go. So we think that it is fair to say that it is the closest we have gotten to AI so far.

22.1   Artificial Neural Networks

It is only too appropriate that we would use a model of a neural network to study the brain. A deep learning algorithm is composed of individual functional units that are connected to each other to simulate a kind of simplified neural network. Just like real neurons, each unit in the network receives inputs—either from outside the network, or from other units in the network—and has outputs, into other neurons or as the final output of the network. Let's look at a diagram of a very simple artificial neural network:

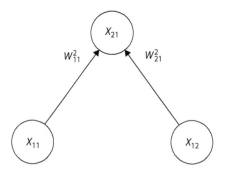

The circles denote nodes and the lines denote direct connections between nodes. The activity of the nodes at the bottom (denoted by X_{11} and X_{12}) is set by two different inputs to the network, and the activity of the node at the top (denoted by X_{21}) is the output of the network. The input layer is similar to the X matrix that you saw in previous chapters, with different rows of the matrix corresponding to different instances of the input. In this network, X_{21} is akin to the model predictions in the previous chapters. As an example, X_{11} might be the weight of an animal, X_{21} might be the wingspan of the animal, and X_{21} might be the model output, which is the predicted airspeed of the animal. This is the familiar regression setting that you saw in previous chapters with two inputs and one output.

In the notation we will use here, the first subscript indicates the layer number and the second subscript indicates the number of the particular unit within the layer. So, X_{12} is the second unit in the first layer, and X_{21} is the first unit in the second layer.

The activity in each unit in the input layer affects the activity in the output unit through a weight: W_{11}^2 and W_{21}^2. The notation for the weights is that the superscript tells us which layer these weights go into, the first subscript tells us the identity of the source unit, and the second subscript tells us the identity of the target unit. For example, W_{11}^2 is a weight from unit one in the first layer to unit one in the second layer and W_{21}^2 is the weight from unit two of the first layer to unit one of the second layer.

How do the weights affect the activity in the second layer? In the simplest case: $X_{21} = W_{11}^2 X_{11} + W_{21}^2 X_{12}$. That is, the activity in the second layer is the weighted linear sum of the activities in the first layer, where the activity in each unit is multiplied by the weight between that unit and the output.

22.1.1 Implementing a Neural Network

Written out in code, this might look something like the following. We use a notation very similar to the one we used in writing out the math, where X_{11} is written as x_11 and W_{11}^2 is written as w_2_11.

```
x11 = 100
x12 = 40
w_2_11 = -1
w_2_21 = 3
x21 = w_2_11 * x11 + w_2_21 * x12
print(x21)
```

```
20
```

In our previous example, this might mean that the airspeed of an animal that weighs 100 g and has a wingspan of 40 cm is 20 km/h (try out different numbers to see how speed is affected by these two factors, given these weights).

Neural networks become more sophisticated when we add more layers to them. For example, here is a three-layer network. There are two units in the input layer, two units in the hidden layer, and one unit in the output layer.

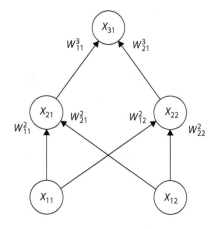

Exercise

Implement the code that calculates the output of this network, using the same notation that we used before. Do not worry about the values of the weights; just make something up.

22.1.2   Activation Functions

Another thing that adds power to neural networks is the addition of nonlinear *activation functions* to the units. Schematically, it would look like this:

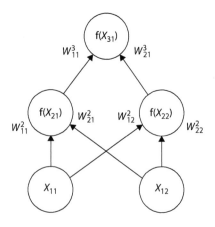

In this case, $f()$ is the activation function. In each unit, this is applied to the weighted sum of the inputs. Often, a nonlinear activation function is used to give the units a threshold of activity that would stimulate them to a response, just like a biological neuron! For example, one typical function to use would be a hyperbolic tangent, which changes very little for very small or very large values, and changes gradually in between. In this case, the output of the first unit in the hidden layer is $\tanh(W_{11}^2 X_{11} + W_{12}^2)$. Another nonlinear activation function that is often used in practice is the rectified linear unit, or ReLU. This function sets the activations that are negative to zero and passes through activations that are positive as they were. We can write this function in Python in very few lines of code

```python
def relu(x):
    if x < 0:
        return 0
    else:
        return x
```

and incorporate this into our two-layer network. In this case, adding the ReLU did not change the output of the network. This is because the output was positive. But you can try changing the weights to see what happens when the weighted sum $W_{11}^2 X_{11} + W_{12}^2$ becomes negative.

```python
x11 = 100
x12 = 40
w_2_11 = -1
w_2_21 = 3
x21 = relu(w_2_11 * x11 + w_2_21 * x12)
print(x21)
```

```
20
```

Exercise

Add the ReLU function as an activation to the hidden and the output layer of the three-layer network that you previously implemented.

22.2 Learning through Gradient Descent and Backpropagation

How do we know what values to assign to the weights of a neural network so that the input-output relationships are accurate? This is the goal of learning with these algorithms. This is implemented using two main ideas. The first idea is that of *gradient descent*. This mechanism for learning is similar to the optimization procedures that we used to fit machine learning models earlier in this chapter, or that we used to register images to each other in chapter 16. It usually starts with some (often random!) guess of what the values of the weights should be and then, based on observations of training data, it determines a small change that should be introduced to improve the accuracy of the input-output relationship: the difference between the true value of *y* for each training sample, and the predicted value of *y*, given the current values of the weights. In each round of training, the weights are adjusted gradually, based on the *gradient* of the weights with respect to the error. The gradient refers to the overall set of changes to the weights that improve the error and we *descend* in the direction of less and less error, which is why this algorithm is known as *gradient descent*. But how does it know how to change the weights to improve the predictions? That is done through *backpropagation*. This is a process that determines the direction and amount of change to apply to each weight based on the errors observed in the current training data.

Backpropagation uses three quantities to determine an error that needs to be adjusted for every weight:

1. The activation of the node feeding into the weight.
2. The slope of the activation function of the node the weight feeds into.
3. The error of the node a weight feeds into.

The first should be straightforward to look up because it should be one of the variables that describe our network. We will come back to the second one; this is also often straightforward (and particularly straightforward for the ReLU activation functions, as we will see). But how do we determine the third of these? This is done by propagating errors back from the output unit. We can determine the error for the output unit, because we are training the network on pairs of inputs and outputs (e.g., animals for which we have measured weight, wingspan, and airspeed). If we want to apply the least-squared criterion in training, this might be the squared error: the squared difference between the output of the network (with the current weight setting) and the true value of the airspeed for this animal. Once we have determined this error, we can send it back to the nodes in the hidden layer that are sending inputs to the output layer. Multiplied by the slope of the activation function in the unit from which the error was backpropagated, and the degree of activation of the node feeding into the weight, this tells us the error for the particular weight we are currently interested in. And once we determine that, we can send this error back to units further down in the network.

Coming back to the slope of the ReLU activation function, we notice that when the input to the function is negative, the output of the ReLU is always zero. This means that for negative inputs, the slope is zero (i.e., the function does not change with changes in the input). For values above zero, the ReLU is the identity function ($f(x) = x$), so its slope is equal to one. This means that we can implement a simple function that calculates the slope of the ReLU function for any input value that is provided:

```
def d_relu(x):
    if x > 0:
        return 1
    else:
        return 0
```

22.2.1 Implementing Backpropagation

As an example of learning, let's imagine a simple data set where the weight and wingspan have been measured in a single animal and we use just that animal's data to train the two-layer network that you saw previously. The weights are initialized as before, so the result, as before, is twenty.

```
x11 = 100
x12 = 40
w_2_11 = -1
w_2_21 = 3
x21 = relu(w_2_11 * x11 + w_2_21 * x12)
print(x21)
```

```
20
```

But let's now consider the situation where this is an incorrect output. That is, the measured airspeed was 15 km/h, and not 20 km/h. In that case, the squared error for the output would be:

```
e21 = (x21 - 15)**2
```

Now, let's propagate this error to determine how the two weights coming into the output unit should change. First, let's calculate the change that should be applied to W_{11}^2. The three components are:

1. The activation of the node feeding into the weight: `x11`.
2. The slope of the activation function of the node the weight feeds into: `d_relu(x21)`.
3. The error of the node a weight feeds into: `e21`.

This means that the change or error for this weight is the product of these three variables:

```
e_2_11 = x11 * d_relu(x21) * e21
```

We can calculate the same for the other weight. This would be the same as the previous example, except we use the activation in the other unit in the input layer:

```
e_2_21 = x12 * d_relu(x21) * e21
print(e_2_21)
print(e_2_11)
```

```
1000
2500
```

These numbers seem rather large, and it is quite common not to apply the full change to the weight, but instead, use a small *learning rate* to make changes to the weights.

```
lr = 10e-6

w_2_11 = w_2_11 - lr * e_2_11
w_2_21 = w_2_21 - lr * e_2_21
```

Having updated the weights, we run the network through again.

```
x21 = relu(w_2_11 * x11 + w_2_21 * x12)
print(x21)
```

```
17.100000000000023
```

We can see that the value that the network is predicting is now closer to the true value. We can repeat the process for as many rounds as we need until the error is deemed small enough (e.g., here we set the threshold at 0.01). In this case, we have to update the weights eighty-one times until the result converges to a value close enough to the true value to be considered close enough.

```
iterations = 1
while e21 > 0.01:
    iterations = iterations + 1
    e21 = (x21 - 15)**2

    e_2_11 = x11 * d_relu(x21) * e21
    e_2_21 = x12 * d_relu(x21) * e21

    w_2_11 = w_2_11 - lr * e_2_11
    w_2_21 = w_2_21 - lr * e_2_21

    x21 = relu(w_2_11 * x11 + w_2_21 * x12)

print(x21)
print(iterations)
```

```
15.098741304117311
81
```

Exercise

Implement the code that calculates the changes to the weights in the three-layer network that you implemented.

22.3 Introducing Keras

Understanding the basic mechanics of training a neural network is useful, but fortunately, when we work with artificial neural networks, we do not have to program all of these low-level operations ourselves. While Scikit-learn only has a limited collection of artificial neural network algorithms, there are several other widely used open-source software libraries for deep learning. Here, we will use the TensorFlow[1] software library, which was originally developed at Google as a tool for their machine learning applications and was made publicly available in 2015. As you have seen throughout this book, having a well-designed application programming interface (API) is very helpful in applying theoretical ideas in practice. Within the TensorFlow library, we will use a particular API called Keras,[2] which makes access to the underlying machinery—constructing networks, training them with data, and using them for prediction—much easier. From TensorFlow, we import the Keras API, and some additional components of the Keras library, a submodule that implements layers of networks, a submodule that implements different model architectures, and a couple of utility functions.

```
from tensorflow import keras
from tensorflow.keras import layers
from tensorflow.keras import models
from tensorflow.keras.utils import to_categorical, set_random_seed
```

22.3.1   A Simple Example

We can start with a small toy example before we proceed to more interesting applications. As you saw in chapter 17, Scikit-learn includes scripts that create synthetic data that we can use to hone our intuitions. We will create a data set like the one we did when we first introduced the idea of classification in section 17.3.2. Using the `make_blobs` function, we create a data set that has two classes ($y=0$ and $y=1$) that are different in terms of their X values (each sample has two features, and the X array has two columns). Just like we did in previous chapters, we split the data into training and test sets. To make sure that the results are identical in each run, we set the random seed with the Keras utility function `set_random_seed`.

---

1. https://www.tensorflow.org/
2. https://keras.io/

```
set_random_seed(2222)
from sklearn.datasets import make_blobs
X, y = make_blobs(centers=3)
from sklearn.model_selection import train_test_split
X_train, X_test, y_train, y_test = train_test_split(X, y, train_size=0.5)
```

Our goal, for now, will be to create a neural network that learns how to classify y values based on X values; recall from chapter 17 that the `make_blobs` function creates three different object categories. The kinds of neural network models we saw earlier are called sequential models in Keras. That is because the activity proceeds sequentially through the network from the bottom to the top (in our diagrams). One way to construct this kind of network is to start by initializing an empty `Sequential()` object and then using the `add` method of this object to add layers. In our first example, we will add layers that are instances of the `Dense` class. This means that there is one weight between each unit in this layer and the previous layer (i.e., they are densely connected; we will see other kinds of layers in just a bit). When initializing each layer we need to specify the number of units in this layer (below sixteen in the first layer, thirty-two in the second layer, and two in the last layer). We can also specify an activation function that governs the output of each unit. Here, we will use the ReLU function in each one. The first layer also needs to know something about the size of the input. In this case, the number of features in X. Finally, the last layer is designed to provide information about the preferred output. Because we are doing a classification task here, rather than a regression task, this layer will have three units—one for each class—and the classification output will be read out from the activity of these three units. If the first unit is the most active, we will predict a zero; if the second unit is most active, we will predict a one; and if the third unit is most active, we will predict a two. This readout, with one unit ultimately winning out over the others, is implemented by a function called a `softmax` function.

```
model = keras.Sequential(
    [
        keras.Input(shape=(X_train.shape[-1],)),
        layers.Dense(8, activation="relu"),
        layers.Dense(16, activation="relu"),
        layers.Dense(3, activation="softmax"),
    ]
)
```

After constructing the network, we need to do one more step before we can fit it, which is to compile the network. This is a feature of TensorFlow that allows it to work fast when fitting even very large models. When we ask to compile a network, TensorFlow figures out a plan for computing the activations in the network, computing the loss, and optimizing it based on the information we provide to it, so after we do this step, the following steps proceed at high speed. When compiling, we can ask for a particular error function to be used for backpropagation relative to the output of the network (also called a *loss function*).

Here, we use the categorical cross-entropy function. We will not get into the details of this function, but suffice it to say that it is an error function that usually works well for classification problems. We can also ask for a particular kind of optimizer to be used to figure out how to apply changes to the weights (e.g., how to determine the learning rate), and we ask for a particular metric to be reported to us as learning proceeds: the classification accuracy.

```
model.compile(loss='categorical_crossentropy', metrics=['accuracy'])
```

Finally, we are ready to fit the model. Keras expects the y input to be converted into its format for categorical variables. This is an array that has one column for each category in y (three in this case). In each row of this array, one value is set to one, corresponding to the category of that observation. The other columns (two columns in this case) are set to zero. For example, the first element of the training data is from the category 2 (i.e., that sample comes from the third blob), and the categorical variable row is equal to [0, 0, 1]

```
categorical_y_train = to_categorical(y_train)
print(y_train[0])
print(categorical_y_train[0])
```

```
2
[0. 0. 1.]
```

We can set the number of times that the neural network will go over all of the training data during training (which are called epochs) and the number of samples that the network will see in each step in training during each of the epochs (which are called the batch size). During training, Keras will update us with information about the value of the loss, which should progressively decrease. The value of categorical cross-entropy is not so easy to interpret, so we also asked to see the classification accuracy, which is easier to understand: a number between zero and one, with perfect classification equal to one. In this case, purely guessing would result in an accuracy of about one-third.

```
history = model.fit(X_train, categorical_y_train, epochs=5, batch_size=5,
                    verbose=2)
```

```
Epoch 1/5
10/10 - 0s - loss: 1.0496 - accuracy: 0.6200 - 480ms/epoch - 48ms/step
Epoch 2/5
```

```
10/10 - 0s - loss: 0.8900 - accuracy: 0.6200 - 12ms/epoch - 1ms/step
```

```
Epoch 3/5
```

```
10/10 - 0s - loss: 0.7799 - accuracy: 0.6200 - 9ms/epoch - 863us/step
```

```
Epoch 4/5
```

```
10/10 - 0s - loss: 0.6883 - accuracy: 0.6200 - 11ms/epoch - 1ms/step
```

```
Epoch 5/5
```

```
10/10 - 0s - loss: 0.6098 - accuracy: 0.6200 - 12ms/epoch - 1ms/step
```

To predict the categories of y in the test data, we need to convert them back from the Keras categorical format to an array of zeros, ones, and twos using the NumPy argmax function. We can then evaluate the predictions relative to the true values using Scikit-learn's accuracy_score function, which quantifies the quality of classification. We find that this network is not amazingly accurate, but it performs far better than a random guess would.

```
y_predict = model.predict(X_test).argmax(axis=1)
from sklearn.metrics import accuracy_score, balanced_accuracy_score
print(accuracy_score(y_test, y_predict))
```

```
0.66
```

One way to think about the performance of this neural network is to quantify the complexity of the model by counting the number of parameters in the model. Keras models include a method that summarizes the model for us, telling us what layers we have in our model and how many parameters are in the model overall.

```
model.summary()
```

```
Model: "sequential"
_____

 Layer (type)                Output Shape             Param #
 ===============================================================
 dense (Dense)               (None, 8)                24

 dense_1 (Dense)             (None, 16)               144

 dense_2 (Dense)             (None, 3)                51
 ===============================================================

Total params: 219
Trainable params: 219
Non-trainable params: 0
_____
```

Let's see how this number, 219, comes about: the first layer of the model has eight units, each of which receives inputs from three different values (the three columns of X)—so the number of weights is twenty-four. The second layer has sixteen units, each of which receives inputs from each of the eight units in the first layer, so eight times sixteen, which is one hundred twenty-eight. In addition to the weight, each unit in this layer also has an additional parameter, a so-called bias term that can adjust the level of activity in the unit up or down by a fixed amount, so we have 128 + 16, which is 144. For the last layer, we have three units, each representing one of the possible output categories, and each of these receives an input from the middle layer, so sixteen times three, which is forty-eight, plus one bias term for each unit, totaling fifty-one. Putting together 24 + 144 + 51 gives us the 219 total parameters in this model. So, is this model complex? Well, that depends on what we compare it to. It is certainly more complex and has more parameters than some of the models we have previously seen. But it is not as complex as some of the neural networks that are the state of the art in deep learning, which can have several million parameters. Such large complexity is achieved by adding many more layers to the network or adding many more units in each layer.

Exercise

Can you make this network even more accurate on these data? What happens if you add more dense layers? What happens if you add more units in each layer? How does each of these changes affect the number of parameters in the model?

22.4 Convolutional Neural Networks

Let's zoom out from the small toy examples you saw until now. As it turns out, the real power of neural networks (particularly for image processing) was discovered when two technological trends converged: the first is the ability to fit weights for neural networks that have a large number of layers. The *deep* in deep learning comes specifically from the stacking of many many layers in models that started becoming feasible around 2010 and that by 2012 were reaching impressive results in the ImageNet benchmark,[3] in which computer vision algorithms are compared in their ability to recognize thousands of different objects in millions of labeled color photographs. One of the main things that made these models feasible was the growing availability of graphical processing units (GPUs) that were used to process the incoming data and to update the massive number of weights in these very large artificial neural networks. The other technological trend that became important around that time was the availability of massive amounts of data. Of course, we do not want to train any machine learning on just one sample, as we did here, but it turns out that training a deep neural network that accurately recognizes objects (or does other sophisticated tasks) requires extraordinary amounts of data. The ImageNet data set itself was therefore a driver of the development of a lot of these algorithms.

At the same time, a clever set of theoretical ideas also drove the development of deep learning, particularly for computer vision. One particularly important idea was that neural networks could incorporate properties of the biological visual system to better learn how to perform computer vision tasks. In particular, researchers who were working on these algorithms in the late 1980s and early 1990s were aware of the classic results that showed that the mammalian primary visual cortex contains neurons that each have their receptive fields in a particular part of the visual field and that these neurons were each particularly sensitive to a narrow range of orientations and spatial frequencies. Each of these receptive fields operates as a small filter that lets through only image features corresponding to its preference, and you can think of the assembly of neurons with these filter properties as performing a convolution of the image on the retina with a kernel of this orientation and spatial frequency (see chapter 14 to remind yourself what convolutions are). Based on this insight, researchers started designing neural networks for computer vision with units that performed a convolution of the input image. Instead of devoting a weight for each pixel in the input image, these units would have only a very small number of weights to fit (e.g., for a 3×3 convolutional kernel they would have only nine weights each). The properties of these convolutional filters—their orientation and spatial frequency sensitivity, for example—would be determined through learning by backpropagation and gradient descent of the convolutional kernel weights, just like the simple network we saw previously.

22.4.1   Implementing Convolutional Neural Networks

To demonstrate this, we will continue to look at the Autism Brain Imaging Data Exchange II (ABIDE II) data set. In the previous sections, we used features of the data that have been computed using a complicated and sophisticated software pipeline: FreeSurfer. FreeSurfer was developed to incorporate a massive amount of knowledge about the brain

---

3. https://www.image-net.org/

and MRI, and provide an understanding of what parts of the image contain cortical areas, white matter, and subcortical regions, etc. The features represent biologically meaningful properties of the brain, such as cortical thickness. The process of creating features that are useful for use in machine learning is known as *feature engineering* and is an important part of many applications of machine learning. This is often a part of the pipeline that requires a deep understanding and intuition about the scientific domain to which machine learning algorithms are applied. A feature of convolutional neural networks that makes them particularly remarkable is that they often do not require feature engineering to be performed in advance. Instead, the algorithms receive the raw images and learn to represent those features in the images that contribute to the task that the neural network is learning. In this case, we will provide to the algorithm as input some of the ABIDE II brain images themselves and ask: Could we train a convolutional neural network algorithm that can accurately classify participants with autism directly from their brain images? When we previously tried to do that based on the FreeSurfer-derived features, we found that we can classify with a reasonable level of accuracy based on these data. Would a convolutional neural network also be able to learn this classification directly from brain images? For simplicity, we will look at the data from only two of the ABIDE II sites (KKI and OHSU) and only at the midsagittal slice of the T1-weighted images for each of ninety subjects from either of these sites, which we have resampled to a resolution of 50 × 50 pixels.

```
from ndslib import load_data
X, y, subject = load_data("abide2_saggitals")
```

```
X_train, X_test, y_train, y_test = train_test_split(
    X, y, train_size=0.8, random_state=100)
```

Instead of the `Dense` layers we used previously, this model will use `Conv2D` layers, which implement the convolution kernels that we are training here. In addition to the number of units per layer and the activation function used by these units, we also need to specify the size of the kernel. For example, here we will use kernels of 3 × 3 pixels. To read out the class information from the network, we need to flatten the output of the second convolutional layer. That is, we need to turn it from a series of two-dimensional images (the results of the convolution filter operation in each of the units) into a one-dimensional vector of outputs. These are then combined through one last `Dense` layer that provides the output of the network (the class information) through the `softmax` function. Here, there are two classes: participants diagnosed with autism and participants who were not diagnosed with autism; so this readout layer has two units and the `softmax` function selects between them.

```
model = keras.Sequential(
    [
        keras.Input(shape=(50, 50, 1)),
```

```
        layers.Conv2D(8, kernel_size=(3, 3), activation="relu"),
        layers.Conv2D(16, kernel_size=(3, 3), activation="relu"),
        layers.Flatten(),
        layers.Dense(2, activation="softmax"),
    ]
)
```

As before, we compile the model to use the categorical cross-entropy function as its loss function and to report to us about the accuracy of classification.

```
model.compile(loss="categorical_crossentropy", metrics=["accuracy"])
```

Before fitting this model, let's contemplate the complexity of this kind of model by looking at the number of parameters in this model.

```
model.summary()
```

```
Model: "sequential_1"
_____

Layer (type)                 Output Shape              Param #
===============================================================
conv2d (Conv2D)              (None, 48, 48, 8)         80

conv2d_1 (Conv2D)            (None, 46, 46, 16)        1168

flatten (Flatten)            (None, 33856)             0

dense_3 (Dense)              (None, 2)                 67714
===============================================================
```

```
Total params: 68,962
```

```
Trainable params: 68,962
```

```
Non-trainable params: 0
```

```
_____
```

That is a lot of parameters! To understand where all the parameters come from in our model, consider that each convolutional filter in the first layer has 3×3 pixels in their kernels, and each one of them also has a bias term, so we have eight times

(9+1), which is eighty. This is more parameters than we had in the first layer of our fully connected network, but consider the dimensionality of the input in this case: each image has 50×50 pixels, so the inputs are 2,500-dimensional, rather than the three-dimensional input in the fully connected network we had before. In the second convolutional layer, each unit has to learn a separate set of weights for each output of the first layer; notice that the output shape is 48×48. This is because, in contrast to the convolution example you saw in chapter 14, the images are not padded with zeros around the edges, so the convolution peels off one layer of pixels around the edges of the image. In this case, that might not matter much. The second layer performs convolutions on these outputs and since we have sixteen units, this means sixteen (units in the second layer) times eight (units in the first layer) times nine (weights in each kernel), adding sixteen bias terms (one for each unit in the second layer), we end up with 1,168 parameters at this layer. Up until now, this does not seem too bad, but before we can read out the output of the network, we have to flatten the output of the second convolutional layer and feed that into the dense layer at the top of the network. At this point, we do have to include a weight for each pixel in the output, so $46 \times 46 \times 16$ inputs into each of the two units at the top layer, which adds up to the 67,714 parameters in that layer. Altogether, 68,962 parameters for the full network. Is this enough to qualify as a deep model? Well, it is all relative. The largest networks that are being developed as we write these lines have billions of parameters, so there is room to grow from here. But these very large models also require much larger training data sets. How well does this network perform this task? We start by fitting it to the data.

```
history = model.fit(X_train, to_categorical(y_train), batch_size=5, epochs=5,
                    verbose=2)
```

```
Epoch 1/5
29/29 - 1s - loss: 12.3917 - accuracy: 0.5845 - 548ms/epoch - 19ms/step
Epoch 2/5
29/29 - 0s - loss: 0.5865 - accuracy: 0.8028 - 147ms/epoch - 5ms/step
Epoch 3/5
29/29 - 0s - loss: 0.3195 - accuracy: 0.9085 - 113ms/epoch - 4ms/step
Epoch 4/5
29/29 - 0s - loss: 0.1622 - accuracy: 0.9507 - 160ms/epoch - 6ms/step
Epoch 5/5
29/29 - 0s - loss: 0.0866 - accuracy: 0.9789 - 146ms/epoch - 5ms/step
```

The fit accuracy seems rather high, reaching almost 100% correct within the five epochs of training. But when we apply the model to the test data, we see much less accurate performance:

```
model.evaluate(X_test, to_categorical(y_test), verbose=2)
```

```
2/2 - 0s - loss: 1.9472 - accuracy: 0.5833 - 140ms/epoch - 70ms/step
[1.9472274780273438, 0.5833333134651184]
```

Looks like the neural network is overfitting to the training data, achieving rather high accuracy of classification in the training data, but then falling to a much lower level of accuracy on the test data. One way to think about this is that the model has so many parameters, and the training data are so small, that the neural network can simply memorize every one of the samples in the training data. Then, when it sees an unfamiliar sample in the test set, it cannot do much better than guess what group this sample belongs to.

What can we do to improve the situation? The convolutional neural network toolbox contains several kinds of tricks to improve model fitting and make it less prone to overfitting. One commonly used trick is called *dropout*. In this approach that was invented by Nitish Srivastava and colleagues [Srivastava et al. 2014], overfitting is prevented by training only some of the units in a particular layer in each batch. That is, some proportion of the units in the layer is dropped out in each batch so that it is not affected by the images in this batch. This means that the units in the layer to which dropout is applied become a little bit less uniformly trained and can respond to different kinds of inputs. This works effectively as a regularization method (you will recall that we previously discussed regularization in Section 21.2.1).

Another way to deal with the high flexibility of the model is to do something to reduce the large number of parameters that the model has. We can do that by reducing the number of convolutional filters in each layer, but another way to do that without hurting the ability of the network to generalize too much is to pool information across neighboring pixels. A common way to do that is to replace every group of pixels with the maximal value of this group of pixels. For example, if we perform a pooling with a 2×2 pixel window on the outputs of a convolutional layer, we would replace every four pixels in the output with a single pixel that has the largest value among the four. This means that the next layer will have much fewer inputs to consider, and hence, much fewer parameters.

```
model = keras.Sequential(
    [
        keras.Input(shape=(50, 50, 1)),
        layers.Conv2D(16, kernel_size=(3, 3), activation="relu"),
        layers.MaxPooling2D(pool_size=(2, 2)),
        layers.Conv2D(32, kernel_size=(3, 3), activation="relu"),
        layers.MaxPooling2D(pool_size=(2, 2)),
        layers.Flatten(),
        layers.Dropout(0.5),
        layers.Dense(2, activation="softmax"),
    ]
)
```

One of the important effects of adding max spatial pooling into the neural network is that the number of parameters goes down. Here, we go from a neural network with almost

70,000 parameters to a network with just over 12,500 parameters. The effect, particularly in this regime with few data samples, is that the network does not fit the training data as well. At the same time, performance on the test data is improved.

```
model.summary()
```

```
Model: "sequential_2"
_____
 Layer (type)                Output Shape              Param #
=================================================================
 conv2d_2 (Conv2D)           (None, 48, 48, 16)        160

 max_pooling2d (MaxPooling2D)  (None, 24, 24, 16)      0


 conv2d_3 (Conv2D)           (None, 22, 22, 32)        4640

 max_pooling2d_1 (MaxPooling  (None, 11, 11, 32)       0
 2D)

 flatten_1 (Flatten)         (None, 3872)              0

 dropout (Dropout)           (None, 3872)              0

 dense_4 (Dense)             (None, 2)                 7746
=================================================================
Total params: 12,546
Trainable params: 12,546
Non-trainable params: 0
_____
```

```
model.compile(loss="categorical_crossentropy", metrics=["accuracy"])
```

```
history = model.fit(X_train, to_categorical(y_train), batch_size=5, epochs=5,
                    verbose=2)
```

```
Epoch 1/5
29/29 - 1s - loss: 2.8656 - accuracy: 0.5563 - 587ms/epoch - 20ms/step
Epoch 2/5
29/29 - 0s - loss: 0.9073 - accuracy: 0.6549 - 227ms/epoch - 8ms/step
Epoch 3/5
29/29 - 0s - loss: 0.6686 - accuracy: 0.6972 - 169ms/epoch - 6ms/step
Epoch 4/5
29/29 - 0s - loss: 0.5708 - accuracy: 0.7465 - 147ms/epoch - 5ms/step
Epoch 5/5
29/29 - 0s - loss: 0.4733 - accuracy: 0.7746 - 131ms/epoch - 5ms/step
```

```
model.evaluate(X_test, to_categorical(y_test), verbose=2)
```

```
2/2 - 0s - loss: 0.7002 - accuracy: 0.6667 - 115ms/epoch - 57ms/step
[0.7002445459365845, 0.6666666865348816]
```

What should we make of this 67% correct performance on the test set? Probably not too much. In addition to issues of class imbalance that we covered previously, one of the concerns with models with this many parameters and multiple layers is that it is hard to say what exactly about the images is the thing that accounts for their accurate level of performance. Here again, we might worry that subtle differences in the images acquired at each site, coupled with different frequencies of each class of subject in each of the sites provide enough clues for the algorithm to substantially exceed 50% performance without actually learning anything truly generalizable about differences between the brains of individuals diagnosed with autism and brains of individuals not diagnosed with autism.

Exercise

Using the subjects variable that was downloaded together with the X and y variables, and merging with the ABIDE II table that you can read with the data=load _data('abide2') command, compute which subjects are from the KKI site and which are from the OHSU site. Using this knowledge, calculate what chance performance would be for an algorithm that relies entirely on identifying the scan site to classify the brains of subjects diagnosed with autism from the brains of subjects not diagnosed with autism. Does this explain the results we obtained with the neural network? How well can the deep learning algorithm classify which site the image was collected in?

22.4.2   Deep learning interpretability

A criticism that has been leveled against the use of machine learning in general and the use of deep learning, in particular, is that the models can be rather inscrutable and are very sensitive to features of the data that are confounded with the variables of interest, such as study site and prevalence of autism in the ABIDE II data set. One way to deal with this criticism is to be very careful in designing our experiments and to scrutinize the data that is used with a very critical eye.

In addition to these measures, researchers in the field of machine learning are developing methods that are focused specifically on trying to understand what neural networks and other machine learning algorithms are doing inside of the black box. These analysis approaches attempt to deconstruct the operations of the neural network by discovering the features of individual images that lead to a particular classification decision. One example is the features of samples in our data set that lead to a classification of a particular image as the brain of an individual diagnosed with autism. This field is broadly known as *interpretable machine learning*, and includes methods that are applied to neural networks as well as to other complex machine learning algorithms, such as the tree algorithms that we demonstrated previously. In some cases, these algorithms can help uncover confounds, like if the deep learning network were to focus on some task-irrelevant features of the image that tell us which site the image is taken from. In other cases, these approaches help uncover interesting relationships in the data. This idea of *interpretable machine learning* brings us full circle to where this part of the book started, with the tension between *prediction* and *explanation*. These methods try to create a bridge between these two goals, by unpacking the reasons that a model accurately predicts.

However, despite these efforts, the criticisms about model interpretability compound with some of the issues that we mentioned in the introduction. Overall, the methods introduced in this book, and particularly in this last chapter need to be approached with care and caution. There are some obvious and pernicious dangers to drawing inferences from large data sets using abstract tools. For example, the performance of many of these models tends to reflect the biases that exist in the training data. Unfortunately, much of the data that we have to train these algorithms to answer questions about individual differences is polluted by the myriad biases that affect how our society treats different individuals. Examples from deep-learning-based technologies such as automated speech recognition [Koenecke et al. 2020] and automated facial recognition [Krishnapriya et al. 2020] demonstrate how these technologies are systematically less accurate when used on members of underrepresented groups. Similarly, complex multivariate analysis of neuroimaging data is also less accurate when trained on a primarily white American sample and then applied to Black Americans [Li et al. 2022]. These results should be a sobering reminder of the kinds of dangers that we run when we fail to account for the circumstances and context in which data are acquired and analyzed. Without serious consideration of potential sources of bias and confusion these methods serve as a double-edged sword that can do more harm than good.

22.5 Additional Resources

The Deep Learning Book[4] by Ian Goodfellow, Yoshua Bengio, and Aaron Courville provides a much deeper introduction to the topic than we can provide.

Keras and TensorFlow are just one of the options that are currently available to implement deep learning methods. Another popular option is Pytorch,[5] an open-source software library originally developed by Facebook/Meta AI.

There are multiple resources to learn more about interpretable machine learning. A good beginner's introduction is provided through a free online book[6] written by Christoph Molnar.

---

4. https://www.deeplearningbook.org/
5. https://pytorch.org/
6. https://christophm.github.io/interpretable-ml-book/

PART VII
Appendices

Appendix 1: Solutions to Exercises

A1.1 The Data Science Toolbox

A1.1.1 Chapter 2

SOLUTION FOR SECTION 2.1.1

```
$ touch new_file.txt

$ touch ~/Documents/new_file.txt

$ mv ~/new_file.txt ~/Documents/new_file.txt
```

The main difference between cp and mv is that after running cp a copy of the file would still exist in the original location, while after running mv there would be no copy remaining.

A1.2 Programming

A1.2.1 Chapter 5

SOLUTIONS FOR SECTION 5.2.3

```
print(type(number_of_subjects * number_of_timepoints))
print(type(number_of_timepoints / number_of_scans))
```

Python changes the results of division into a float because results of some division operations between integers are not integers themselves.

```
my_string = "supercalifragilisticexpialidocious"
number_of_li = my_string.count("li")
print(number_of_li)
```

```
print(1 == True)
print(0 == False)
```

SOLUTIONS SECTION 5.3.1

```
print(random_stuff[-1])
print(random_stuff[-4])
print(random_stuff[-len(random_stuff)])
```

```
list1 = [1, 2, 3]
list2 = [4, 5, 6]

list2.reverse()
list3 = list2[:-1] + list1
print(list3)
```

SOLUTION FOR SECTION 5.3.2

```
fruit_prices["pear"] = [3, 4, 5]
print(fruit_prices["pear"][1])
```

SOLUTION FOR SECTION 5.4.1

```
my_int = 6
print(type(my_int))
print(dir(my_int))
my_float = 6.0
print(type(my_float))
print(dir(my_float))
```

SOLUTION FOR SECTION 5.5.1

```
if mango < 0.5:
    print("Mangoes are super cheap; get a bunch of them!")
elif mango < 1.0:
    print("Get one mango from the store.")
elif mango > 2.0 and mango < 5.0:
    print("Mangos are expensive. Maybe just one")
else:
    print("Meh. I don't really even like mangoes.")
```

SOLUTION FOR SECTION 5.5.2

```
num_elems = len(random_stuff)

for i in range(num_elems):
    val = random_stuff[i]
    print(f"{i}, {i**2}")
```

SOLUTION FOR SECTION 5.5.4

```
list_with_squares = [(ii, ii ** 2) for ii in range(len(random_stuff))]
print(list_with_squares)
```

SOLUTION FOR SECTION 5.5.5

```
num = 800

if num > 500:
    if num < 900:
        if num > 700:
            print("Great number.")
        else:
            print("Terrible number.")
```

```
num = 800

if num > 500:
    if num < 900:
    if num > 700:
            print("Great number.")
        else:
            print("Terrible number.")
```

SOLUTION FOR SECTION 5.7.1

```
def create_sample(x, mu, sd, n):
    result = []
    for ii in range(n):
        result.append(add_noise(x, mu, sd))
    return result

sample = create_sample(4, 1, 2, 10)
print(sample)
```

SOLUTION FOR SECTION 5.7.2

```
a_list = [1,2,3,4,5]
arg_printer(min, a_list)
arg_printer(len, a_list)
```

SOLUTION FOR SECTION 5.8.4

```python
from math import pi

class Circle:

    def __init__(self, radius):
        self.radius = radius

    def area(self):
        return pi * self.radius**2

    def circumference(self):
        return 2 * pi * self.radius

class Square:
    def __init__(self, side):
        self.side = side

    def area(self):
        return self.side ** 2

    def circumference(self):
        return self.side * 4
```

SOLUTION FOR SECTION 5.8.8

```python
class Square:
    def __init__(self, side):
        self.side = side

    def area(self):
        return self.side ** 2

    def circumference(self):
        return self.side * 4

    def __mul__(self, prey):
        new_area = self.area() + prey.area()
        self.side = sqrt(new_area)

s1 = Square(4)
s2 = Square(2)
print(s1.side)
print(s1.area())

print("Eat the other square")
s1 * s2

print(s1.side)
print(s1.area())
```

A1.3 Scientific Computing

A1.3.1 Chapter 8

SOLUTIONS SECTION 8.2.3

```
list_of_list_of_lists = [[[1, 1, 2, 3, 5], [8, 13, 21, 34, 55]],
                         [[2, 2, 4, 6, 10], [16, 26, 42, 68, 55]]]

array3d = np.array(list_of_list_of_lists)
print(array3d)
print(array3d[1, 0, 2])
```

The first item in the strides for the first array represents the ten items in the second dimension of the array that you would have to pass through to get to the first item in the second row (times 8 bytes per item). The first item in the strides for the second array represents the four items (in the second dimension) times eight items (in the third dimension) that you would to pass in order to get to the first item in the item along each of these dimensions (times 8 bytes per item).

SOLUTION FOR SECTION 8.2.8

```
bold.std()
bold[(bold >=1) & (bold<100)].std()
```

SOLUTION FOR SECTION 8.2.9

```
solution1 = np.arange(1, 100, 2)
solution2 = np.ones((3, 5))
array_with_numbers = np.arange(150)
solution3 = (array_with_numbers > 0) & (array_with_numbers <= 100)
solution4 = np.mod(array_with_numbers, 3) == 0
solution5 = np.random.random_sample(100) * 10
```

SOLUTION FOR SECTION 8.2.10

```
import scipy.interpolate as spi
x = np.arange(0, 360, 2)
interp = spi.interp1d(x, bold[32, 32, 12])
new_x = np.arange(0, 358, 0.5)
high_res = interp(new_x)
```

A1.3.2 Chapter 9

SOLUTION FOR SECTION 9.3.1

```
subjects["IQ_sub_diff"] = subjects["IQ_Vocab"] - subjects["IQ_Matrix"]
```

SOLUTION FOR SECTION 9.3.2

```
new_df = subjects[subjects["IQ"].notnull()]
subjects.dropna(inplace=True)
```

SOLUTION FOR SECTION 9.3.4

```
# One way of doing this:
subjects.groupby(["Handedness", "Gender", "age_less_than_10"]).mean()

# Another way of doing this:
subjects[(subjects["Handedness"] == "Right") &
         (subjects["Gender"] == "Male") &
         (~subjects["age_less_than_10"])].mean()
```

SOLUTION FOR SECTION 9.4.1

```
# Part 1
gender_groups_mean = joined.groupby(["Gender", "tractID", "nodeID"]).mean()
male_means = gender_groups_mean.loc[("Male", "Left Cingulum Cingulate")]
female_means = gender_groups_mean.loc[("Female", "Left Cingulum Cingulate")]

fig, ax = plt.subplots()
ax.plot(male_means["fa"])
ax.plot(female_means["fa"])
ax.set_xlabel("Node")
ax.set_ylabel("Fractional anisotropy")

# Part 2

for tract in joined["tractID"].unique():
    fig, ax = plt.subplots()
    ax.plot(group_means.loc[(False, tract)]["fa"])
    ax.plot(group_means.loc[(True, tract)]["fa"])
    ax.set_xlabel("Node")
    ax.set_ylabel("Fractional anisotropy")
```

A1.3.3   Chapter 10

SOLUTION FOR SECTION 10.1.1

```
fig, ax = plt.subplots()
ax.plot(trial,
        first_block,
        marker='o',
        linestyle='--',
        label="First block",
        color="lightseagreen")

ax.plot(trial,
        middle_block,
        marker='v',
        linestyle='--',
```

```
        label="Middle block",
        color="palegoldenrod")

ax.plot(trial,
        last_block,
        marker='^',
        linestyle='--',
        label="Last block",
        color="firebrick")

ax.set_xlabel("Trials")
ax.set_ylabel("Percent correct")
ax.legend()
title = ax.set_title("Harlow, 1949")
```

SOLUTION FOR SECTION 10.1.2

```
fig, ax = plt.subplots(5, 4)
min_fa = min([younger_fa.min(), older_fa.min()])
max_fa = max([younger_fa.max(), older_fa.max()])

for tract_idx in range(20):
    pathway = tracts[tract_idx]
    ax.flat[tract_idx].plot(younger_fa[pathway], color="C0")
    ax.flat[tract_idx].plot(older_fa[pathway], color="C1")
    ax.flat[tract_idx].set_ylim([min_fa, max_fa])
    ax.flat[tract_idx].set_title(pathway)
    ax.flat[tract_idx].axis("off")


fig.set_tight_layout("tight")
fig.set_size_inches([10, 8])

ax[4, 0].set_xlabel("Node")
label = ax[4, 0].set_ylabel("Fractional Anisotropy")
```

SOLUTION FOR SECTION 10.3.1

```
max_bold = bold.max()
min_bold = bold.min()
fig, ax = plt.subplots(3, 6)
for frame_idx in range(18):
    ax.flat[frame_idx].matshow(
        bold[:, :, 10,
        frame_idx * 10],
        cmap="bone",
        vmax=max_bold,
        vmin=min_bold)
    ax.flat[frame_idx].axis("off")
fig.set_tight_layout("tight")
```

A1.4   Neuroimaging in Python

A1.4.1   Chapter 11

SOLUTION FOR SECTION 11.2.2

```
layout = BIDSLayout("path/to/bids-examples/ds011")
print(layout)
bold_files = layout.get(suffix="bold", return_type='filename', extension="nii.gz")
trs = [layout.get_metadata(ff)["RepetitionTime"] for ff in bold_files]
print(trs)
```

A1.4.2   Chapter 12

SOLUTIONS FOR SECTION 12.1

1. The results are different, but only by a very small amount. In fact, they are identical if voxels with standard deviation smaller than 0.001 are excluded from the final calculation:

```
mean = np.mean(data_bold, -1)
std = np.std(data, -1)
tsnr_numpy = np.zeros(mean.shape)
std_idx = std > 1.0e-3
tsnr_numpy[std_idx] = mean[std_idx] / std[std_idx])
```

2. MRIQC signal-to-noise ratio (SNR) is larger than the one calculated before. One reason this could be is that it is correcting the data for motion-related artifacts before calculating SNR.

A1.4.3   Chapter 13

SOLUTIONS SECTION 13.1

1. This is $[-1, 1, 1]$
2. The first dimension increases from front to back (P), the second dimension increases from inferior to superior (S) the last dimension increases either from right-to-left (L) or from left-to-right (R), so it is either LPS or RPS.
3. $C_{2,1} = A_{2,1}B_{1,1} + A_{2,2}B_{2,1} + A_{2,3}B_{3,1} = 4 \cdot 7 + 5 \cdot 9 + 6 \cdot 11 = 139$
   $C_{2,2} = A_{2,1}B_{1,2} + A_{2,2}B_{2,2} + A_{2,3}B_{3,2} = 4 \cdot 8 + 5 \cdot 10 + 6 \cdot 12 = 154$

```
def matmul(a, b):
    result = np.zeros((a.shape[0], b.shape[1]))
    for ii in range(result.shape[0]):
        for jj in range(result.shape[1]):
            for kk in range(a.shape[1]):
                result[ii, jj] = result[ii, jj] + a[ii, kk] * b[kk, jj]

    return result
```

SOLUTION FOR SECTION 13.3.2

```
print(affine_bold @ np.array([data_bold.shape[0],
                              data_bold.shape[1],
                              data_bold.shape[2], 1]))
```

SOLUTION FOR SECTION 13.3.3

The rotation matrix is defined as:

$$A_\theta = \begin{bmatrix} \cos(\theta) & -\sin(\theta) \\ \sin(\theta) & \cos(\theta) \end{bmatrix}$$

Its inverse is:

$$A_{\theta-1} = \begin{bmatrix} \cos(-\theta) & -\sin(-\theta) \\ \sin(-\theta) & \cos(-\theta) \end{bmatrix}$$

The transpose of this matrix is the matrix where columns and rows have been switched:

$$A_\theta = \begin{bmatrix} \cos(\theta) & \sin(\theta) \\ -\sin(\theta) & \cos(\theta) \end{bmatrix}$$

But $\cos(\theta) = \cos -\theta$ and $-\sin(\theta)$ is $\sin(-\theta)$ so, replacing this in this is also:

$$A_\theta = \begin{bmatrix} \cos(-\theta) & \sin(\theta) \\ \sin(-\theta) & \cos(-\theta) \end{bmatrix}$$

For the same reason, $\sin(\theta) = -\sin(-\theta)$, so replacing that in the first row of the matrix yields the inverse as shown previously.

SOLUTION FOR SECTION 13.3.4

```
# Part 1:
img_bold0 = nib.Nifti1Image(data_bold[:, :, :, 0], img_bold.affine)
img_bold0_resampled = resample_from_to(img_bold0, (img_t1.shape, img_t1.affine))

data_bold0_resampled = img_bold0_resampled.get_fdata()

fig, ax = plt.subplots(1, 2)
ax[0].matshow(data_t1[:, :, data_t1.shape[-1]//2])
im = ax[1].matshow(data_bold0_resampled[:, :, data_bold0_resampled.shape[-1]//2])
```

Part 2: We would expect more loss of information going from the T1-weighted data to the fMRI data than vice versa. However, this is exclusively because of loss of information in the T1-weighted data. There is no information in the fMRI data beyond the resolution at which they were originally sampled.

A1.5 Image Processing

A1.5.1 Chapter 14

SOLUTION FOR SECTION 14.4.1

```
print(gray_img[200, 70])
print(gray_img[400, 200])
```

The values of these pixels are relatively low (less than 0.5). This is because they have high values in the red and blue channels and these are weighted much lower in the conversion to gray scale than the values of the green channel.

SOLUTION FOR SECTION 14.4.2

```
mri_img = np.random.randn(100, 100, 100)

kernel = np.ones((20, 20, 20))
result = np.zeros(mri_img.shape)
padded_mri_img = np.pad(mri_img, int(kernel.shape[0] / 2))

for ii in range(result.shape[0]):
    for jj in range(result.shape[1]):
        for kk in range(result.shape[2]):
            neighborhood = padded_mri_img[ii:ii+kernel.shape[0],
                                          jj:jj+kernel.shape[1],
                                          kk:kk+kernel.shape[2]]
            weighted_neighborhood = neighborhood * kernel
            conv_voxel = np.sum(weighted_neighborhood)
            result[ii, jj, kk] = conv_voxel
```

SOLUTION FOR SECTION 14.4.3

```
from skimage.filters import sobel, sobel_h, sobel_v
fig, ax = plt.subplots(1, 3)
sobel_collins = sobel(gray_img)
ax[0].matshow(sobel_collins)
sobel_v_collins = sobel_v(gray_img)
ax[1].matshow(sobel_v_collins)
sobel_h_collins = sobel_h(gray_img)
ax[2].matshow(sobel_h_collins)
fig, ax = plt.subplots()
sobel_brain = sobel(brain)
ax.matshow(sobel_brain[:, :, 10])
```

These functions are edge detectors! The _h and _v versions of this function emphasize specifically horizontal and vertical edges. These functions do not work on three-dimensional images because they do not have a clear notion of horizontal and vertical associated with them.

SOLUTION FOR SECTION 14.4.4

```
# Part 1
def morphological_white_tophat(image, selem=disk(7)):
    return image - dilation(erosion(image, selem=selem), selem=selem)

def morphological_black_tophat(image, selem=disk(7)):
    return image - erosion(dilation(image, selem=selem), selem=selem)

fig, ax = plt.subplots(1, 2)
ax[0].matshow(morphological_white_tophat(shepp_logan))
ax[1].matshow(morphological_black_tophat(shepp_logan))

fig, ax = plt.subplots(1, 2)
ax[0].matshow(morphological_white_tophat(shepp_logan, selem=disk(14)))
ax[1].matshow(morphological_black_tophat(shepp_logan, selem=disk(14)))


# Part 2
selem = disk(7)
fig, ax = plt.subplots(1, 2)
ax[0].matshow(brain[:, :, 10])
ax[1].matshow(dilation(erosion(brain[:, :, 10], selem=selem), selem=selem))


from skimage.morphology import ball
selem = ball(7)
fig, ax = plt.subplots(1, 2)
ax[0].matshow(brain[:, :, 10])
ax[1].matshow(dilation(erosion(brain, selem=selem), selem=selem)[:, :, 10])
```

A1.5.2   Chapter 15

SOLUTION FOR SECTION 15.1.1

```
# Part 1
min_intraclass_variance = np.inf

unique_vals = np.unique(slice10)

for ii in range(len(unique_vals)):
    for jj in range(ii, len(unique_vals)):
        candidate1 = unique_vals[ii]
        candidate2 = unique_vals[jj]
        foreground1 = slice10[slice10 < candidate1]
        foreground2 = slice10[(slice10>=candidate1) & (slice10 < candidate2)]
        background = slice10[slice10 >= candidate2]
        if len(foreground1) and len(foreground2) and len(background):
            foreground1_variance = np.var(foreground1) * len(foreground1)
            foreground2_variance = np.var(foreground2) * len(foreground2)
            background_variance = np.var(background) * len(background)
            intraclass_variance = (foreground1_variance +
                                   foreground2_variance +
                                   background_variance)
            if intraclass_variance < min_intraclass_variance:
                min_intraclass_variance = intraclass_variance
                threshold1 = candidate1
                threshold2 = candidate2

segmentation = np.zeros_like(slice10)
segmentation[(slice10 > threshold1) & (slice10 < threshold2)] = 1
fig, ax = plt.subplots()
ax.imshow(slice10, cmap="bone")
p = ax.imshow(segmentation, alpha=0.5)


from skimage.filters import threshold_multiotsu
threshold1, threshold2 = threshold_multiotsu(slice10)
```

```
segmentation = np.zeros_like(slice10)
segmentation[(slice10 > threshold1) & (slice10 < threshold2)] = 1
fig, ax = plt.subplots()
ax.imshow(slice10, cmap="bone")
p = ax.imshow(segmentation, alpha=0.5)

# Part 2
from skimage.filters import try_all_threshold
try_all_threshold(slice10)
```

The minimum and Otsu methods both seem to be similarly good at doing this. One way to evaluate this objectively would be to look at the histogram of this slice across all time points in the fMRI time series and seeing that they all remain in a similar segment.

A1.5.3   Chapter 16

SOLUTION FOR SECTION 16.1.2

```
metric = CCMetric(3)
sdr = SymmetricDiffeomorphicRegistration(metric)

mapping = sdr.optimize(mni_data, t1_resamp_data, prealign=affine3d.affine)
t1_warped = mapping.transform(t1_resamp_data)

fig, axes = plt.subplots(1, 3, figsize=(8, 4))
ax = axes.ravel()

ax[0].imshow(mni_data[:, :, 85]/np.max(mni_data))
ax[1].imshow(t1_warped[:, :, 85]/np.max(t1_warped))

stereo = np.zeros((193, 229, 3), dtype=np.uint8)
stereo[..., 0] = 255 * mni_data[:, :, 85]/np.max(mni_data)
stereo[..., 1] = 255 * t1_warped[:, :, 85]/np.max(t1_warped)
ax[2].imshow(stereo)
fig.tight_layout()
```

Diffeomorphic registration substantially improves the registration to the template. There is a risk of warping the features substantially, to a point where they take on an unrealistic shape, similar to some of the distortions that happen in the photo.

A1.6   Machine Learning

A1.6.1   Chapter 18

SOLUTION FOR SECTION 18.2.3

```
# Part 1
from sklearn.metrics import mean_squared_error, mean_absolute_error

print(mean_squared_error(y, y_pred))
print(mean_absolute_error(y, y_pred))
```

```
# Part 2

import numpy as np
def max_absolute_error(y, y_pred):
    return np.max(np.abs(y - y_pred))

print(max_absolute_error(y, y_pred))
```

A1.6.2   Chapter 19

SOLUTION FOR SECTION 19.1.1

```
est = LinearRegression()
est.fit(x[:, None], y)

test_x, test_y = make_xy(30)

x_range = np.linspace(test_x.min(), test_x.max(), 100)
reg_line = est.predict(x_range[:, None])

fig, ax = plt.subplots()
ax.scatter(test_x, test_y)
ax.plot(x_range, reg_line)

mse = mean_squared_error(y, est.predict(test_x[:, None]))
ax.set_title(f"Mean squared error: {mse:.2f}");
```

The error does increase when applied to new data, but this increase is much smaller than the increase observed for the tenth-order polynomial.

A1.6.3   Chapter 20

SOLUTION FOR SECTION 20.1.3

We have higher out-of-sample accuracy because our training sample is larger, so the model has more data to learn from. The downside is that our evaluation data set is smaller, so we get widely varying test accuracies (between 0.45 and 0.62) that depend on the characteristics of the test sample.

SOLUTION FOR SECTION 20.1.5

```
y_adjusted = abide_data['age_resid']
r2_cv = cross_val_score(est, X, y_adjusted, cv=k)

print("Individual fold scores:", r2_cv)
print(f"\nMean cross-validated R^2: ", np.mean(r2_cv))
```

This result tells us that there is not much information about age, above and beyond the information about site.

SOLUTION FOR SECTION 20.2.1

```
n_features = [5, 30, 100]
r2 = []
train_scores
test_scores
fig, axes = plt.subplots(1, 3)
for ii in range(len(feature_sets)):
    ax = axes[ii]
    X_subset = X.sample(n_features[ii], axis=1)
    results = learning_curve(est, X_subset, y, train_sizes=train_sizes,
                             cv=5, shuffle=True)
    sizes, train_scores, test_scores = results
    train_mean = train_scores.mean(1)
    test_mean = test_scores.mean(1)

    ax.plot(sizes, train_mean, 'o-', label='Training')
    ax.plot(sizes, test_mean, 'o-', label='Test');
    ax.grid(True, axis='y')
    ax.set_ylim(0, 1)
    ax.set_xlabel('Sample size (n)')
    ax.set_ylabel('$R^2$');
    ax.set_title(f"{n_features[ii]} features used" )
fig.set_tight_layout("tight")
```

A1.6.4   Chapter 21

SOLUTIONS SECTION 21.2.1

```
# Part 1
n_features = [10, 100, 400]
for nn in n_features:
    X = scale(features.sample(nn, axis=1, random_state=99))
    alpha = np.logspace(-3, 1, 100)
    ax = plot_coef_path(Lasso, X, y, alpha, max_iter=2000)
    ax.set_title(f"{nn} features")
    ax.set_xlim(1e-3, 10)

# Part 2
y_resid = phenotypes['age_resid']
n_features = [10, 100, 400]
for nn in n_features:
    X = scale(features.sample(nn, axis=1, random_state=99))
    ax = plot_coef_path(Lasso, X, y_resid, alpha, max_iter=2000)
    ax.set_title(f"{nn} features")
    ax.set_xlim(1e-3, 10)
```

SOLUTION FOR SECTION 21.2.1

```
from sklearn.model_selection import validation_curve, cross_val_score
from sklearn.linear_model import LinearRegression
from ndslib.viz import plot_train_test

x_range = np.logspace(-5, 5, 100)
train_scores, test_scores = validation_curve(
    Ridge(), X, y, param_name='alpha',
    param_range=x_range, cv=5, scoring='r2')

ols_r2 = cross_val_score(
    LinearRegression(), X, y, scoring='r2', cv=5).mean()

ax = plot_train_test(
    x_range, train_scores, test_scores, 'Ridge',
    hlines={'OLS (test)': ols_r2})

ax.set_title("Age prediction performance")
```

A1.6.5   Chapter 22

SOLUTION FOR SECTION 22.1.1

```
x11 = 100
x12 = 40
w_2_11 = -2
w_2_21 = 3
w_2_12 = 0.5
w_2_22 = -1
x21 = w_2_11 * x11 + w_2_21 * x12
x22 = w_2_12 * x11 + w_2_22 * x12
w_3_11 = 1
w_3_21 = 2
x31 = w_3_11 * x21 + w_3_21 * x22
print(x31)
```

SOLUTION FOR SECTION 22.1.2

```
x11 = 100
x12 = 40
w_2_11 = -2
w_2_21 = 3
w_2_12 = 0.5
w_2_22 = -1
x21 = relu(w_2_11 * x11 + w_2_21 * x12)
x22 = relu(w_2_12 * x11 + w_2_22 * x12)
w_3_11 = 1
w_3_21 = 2
x31 = relu(w_3_11 * x21 + w_3_21 * x22)
print(x31)
```

SOLUTION FOR SECTION 22.2.1

```
x11 = 100
x12 = 40
w_2_11 = -2
w_2_21 = 3
w_2_12 = 0.5
w_2_22 = -1
x21 = relu(w_2_11 * x11 + w_2_21 * x12)
x22 = relu(w_2_12 * x11 + w_2_22 * x12)
w_3_11 = 1
w_3_21 = 2
x31 = relu(w_3_11 * x21 + w_3_21 * x22)
e31 = (x31 - 15)**2

iterations = 0

while e31 > 0.01:
    iterations = iterations + 1
    # This part is similar to the two-layer network
    e_3_11 = x21 * d_relu(x31) * e31
    e_3_21 = x22 * d_relu(x31) * e31

    w_3_11 = w_3_11 - lr * e_3_11
    w_3_21 = w_3_21 - lr * e_3_21

    # Propagate the error to the second layer:
    e21 = w_3_11 * e_3_11
    e22 = w_3_21 * e_3_21

    e_2_11 = x11 * d_relu(x21) * e21
    e_2_21 = x12 * d_relu(x21) * e22
    e_2_12 = x12 * d_relu(x22) * e21
    e_2_22 = x12 * d_relu(x22) * e22

    w_2_11 = w_2_11 - lr * e_2_11
    w_2_21 = w_2_21 - lr * e_2_21
    w_2_12 = w_2_12 - lr * e_2_11
    w_2_22 = w_2_22 - lr * e_2_21

    # Forward propagate through the network again with the same inputs
    x21 = relu(w_2_11 * x11 + w_2_21 * x12)
    x22 = relu(w_2_12 * x11 + w_2_22 * x12)
    x31 = relu(w_3_11 * x21 + w_3_21 * x22)
    e31 = (x31 - 15)**2

print(x31)
print(iterations)
```

SOLUTION FOR SECTION 22.3.1

```
model = keras.Sequential(
    [
        keras.Input(shape=(X_train.shape[-1],)),
        layers.Dense(8, activation="relu"),
        layers.Dense(16, activation="relu"),
        layers.Dense(32, activation="relu"),
        layers.Dense(16, activation="relu"),
        layers.Dense(3, activation="softmax"),
    ]
)

model.compile(loss='categorical_crossentropy', metrics=['accuracy'])
history = model.fit(X_train, categorical_y_train, epochs=5, batch_size=5,
                    verbose=2)

y_predict = model.predict(X_test).argmax(axis=1)
print(accuracy_score(y_test, y_predict))

model.summary()
```

Increasing the number of layers and units can improve accuracy, but it also increases the number of parameters.

SOLUTION FOR SECTION 22.4.1

```
import pandas as pd
import numpy as np
subject_table = load_data("abide2")
new_table = pd.merge(pd.DataFrame(data=dict(subject=subject)),
                     subject_table,
                     left_on="subject",
                     right_on="subject")
new_table.groupby("site")["group"].value_counts()

aut_in_kki = 22 / (67 + 22)
aut_in_ohsu = 36 / (53 + 36)
guess_with_site = np.mean([1-aut_in_kki, 1-aut_in_ohsu])
print(guess_with_site)

y_site = to_categorical(new_table["site"] == "ABIDEII-KKI_1")
X_train, X_test, y_train, y_test = train_test_split(
    X, y_site, train_size=0.8, random_state=100)
history = model.fit(X_train, y_train, batch_size=5, epochs=5,
                    verbose=2)

model.evaluate(X_test, y_test, verbose=2)
```

Based on these results (note that chance performance given knowledge of site is 0.67 and accuracy of classifying the site is very high) we would have to conclude that the performance of the algorithm could be explained entirely by classifying the site.

Appendix 2: `ndslib` Function Reference

The `ndslib` software library implements functions that we use in the book to download data sets, and to encapsulate some computations. We provide descriptions of the functions for each module subsequently. Additional documentation can be found in the ndslib library website.[1]

ndslib.data.load_data(dataset, fname=None)

Loads data for use in examples.

Parameters:

data set [str] The name of a data set. Can be one of:

bold_numpy : Read a fMRI time series as a numpy array.
bold_volume : Read a single volume of a fMRI time series as a numpy array.
afq : Read diffusion MRI data in tabular format.
age_groups_fa : Read AFQ data and return dataframe divided by age groups.
abide2_saggitals : Read ABIDE II mid saggitals as numpy arrays.
abide2: Read the ABIDE II FreeSurfer features.

fname [str, optional.] If provided, data will be cached to this local path and retrieved from there on future calls with the same value.

ndslib.data.download_bids_dataset()

Makes a minimal Brain Imaging Data Structure (BIDS) data set with one fMRI subject from OpenNeuro data set ds001233.

ndslib.viz.imshow_with_annot(im, vmax=40)

Like imshow, but with added annotation of the array values.

Parameters: **im** : numpy array

ndslib.viz.plot_diffeomorphic_map(mapping, ax, delta=15, direct_grid_shape=None, direct _grid2world=- 1, inverse_grid_shape=None, inverse_grid2world=- 1)

1. https://neuroimaging-data-science.github.io/ndslib/

Draw the effect of warping a regular lattice by a diffeomorphic map. Draws a diffeo-morphic map by showing the effect of the deformation on a regular grid. The resulting figure contains two images: the direct transformation is plotted to the left, and the inverse transformation is plotted to the right.

Parameters:

mapping [DiffeomorphicMap] The diffeomorphic map to be drawn

delta [int, optional] The size (in pixels) of the squares of the regular lattice to be used to plot the warping effects. Each square will be delta x delta pixels. By default, the size will be 10 pixels.

fname [string, optional] The name of the file the figure will be written to. If None (default), the figure will not be saved to disk.

direct_grid_shape [tuple, shape (2,), optional] The shape of the grid image after being deformed by the direct transformation. By default, the shape of the deformed grid is the same as the grid of the displacement field, which is by default equal to the shape of the fixed image. In other words, the resulting deformed grid (deformed by the direct transformation) will normally have the same shape as the fixed image.

direct_grid2world [array, shape (3, 3), optional] The affine transformation map-ping the direct grid's coordinates to physical space. By default, this transformation will correspond to the image-to-world transformation corresponding to the default direct_grid_shape (in general, if users specify a direct_grid_shape, they should also specify direct_grid2world).

inverse_grid_shape [tuple, shape (2,), optional] The shape of the grid image after being deformed by the inverse transformation. By default, the shape of the deformed grid under the inverse transform is the same as the image used as moving when the diffeomorphic map was generated by a registration algorithm (so it corresponds to the effect of warping the static image toward the moving).

inverse_grid2world [array, shape (3, 3), optional] The affine transformation map-ping inverse grid's coordinates to physical space. By default, this transformation will correspond to the image-to-world transformation corresponding to the default inverse_grid_shape (in general, if users specify an inverse_grid_shape, they should also specify inverse_grid2world).

Returns:

warped_forward [array] Image with the grid showing the effect of transform-ing the moving image to the static image. The shape will be direct_grid_shape if specified, otherwise the shape of the static image.

warped_backward [array] Image with the grid showing the effect of transforming the static image to the moving image. Shape will be inverse_grid_shape if specified, otherwise the shape of the moving image.

Notes:

The default value for the affine transformation is "−1" to handle the case in which the user provides None as input meaning identity. If we used None as default, we wouldn't know if the user specifically wants to use the identity (specifically passing None) or if it was left unspecified, meaning to use the appropriate default matrix.

ndslib.viz.plot_coef_path(estimator, X, y, alpha, **kwargs)

Plot the coefficient path for a sklearn estimator.

Parameters:

estimator [sklearn estimator] For example "Lasso()"

X [ndarray (n, m)] Feature matrix

y [ndarray (n,)] Target matrix

Returns: ax : Matplotlib *Axes* object

ndslib.viz.plot_train_test(x_range, train_scores, test_scores, label, hlines=None)

Plot train/test R^2

Parameters:

x_range [sequence] The range of x values used (e.g., number of features, number of samples)

train_scores [sequence] The train r2_score corresponding to different x values

test_scores [sequence] The test r2_score corresponding to different x values

label [str] Used in the legend labels.

hlines [dict] A dictionary where keys are labels and values are y values for hlines.

Returns: ax : Matplotlib *Axes* object.

ndslib.viz.plot_learning_curves(estimators, X_sets, y, train_sizes, labels=None, errors=True, **kwargs)

Generate multipanel plot displaying learning curves for multiple predictor sets and/or estimators.

Parameters:

estimators [list,] A Scikit-learn Estimator or list of estimators. If a list is provided, it must have the same number of elements as X_sets.

X_sets [list,] An NDArray or similar object, or list. If a list is passed, it must have the same number of elements as estimators.

y [ndarray] A one-dimensional numpy array (or Pandas Series) representing the outcome variable to predict.

train_sizes [list] List of ints providing the sample sizes at which to evaluate the estimator.

labels [list, optional.] List of labels for the panels. Must have the same number of elements as X_sets.

errors [bool, optional.] If True, plots error bars representing 1 StDev. Default: True.

kwargs [dict, optional] Optional keyword arguments passed on to sklearn's learning_curve utility.

ndslib.viz.plot_graphviz_tree(tree, feature_names)

Takes a tree as input, calls Scikit-learn's export_graphviz function to generate an image of the tree using graphviz, and then plots the result in-line.

Parameters: tree: sklearn tree object **feature_names** : sequence of strings

ndslib.image.gaussian_kernel(x=20, sigma=4)

Construct a two-dimensional Gaussian kernel for image processing.

Parameters:

x [int, optional] The number of pixels on a side for the filter. Default : 20

sigma [float, optional] The standard deviation parameter for the Gaussian. Default : 4

Returns:

gauss [ndarray] Contains the values of the two-dimensional Gaussian normalized to sum to one.

BIBLIOGRAPHY

[1] B B Avants, C L Epstein, M Grossman, and J C Gee. Symmetric diffeomorphic image registration with cross-correlation: evaluating automated labeling of elderly and neurodegenerative brain. *Med. Image Anal.*, 12(1):26–41, February 2008.

[2] Richard A I Bethlehem, Jakob Seidlitz, Rafael Romero-Garcia, Stavros Trakoshis, Guillaume Dumas, and Michael V Lombardo. A normative modelling approach reveals age-atypical cortical thickness in a subgroup of males with autism spectrum disorder. *Comm. Biol.*, 3(1):1–10, 2020.

[3] Patrick Beukema, Jörn Diedrichsen, and Timothy D Verstynen. Binding during sequence learning does not alter cortical representations of individual actions. *J. Neurosci.*, 39(35):6968–6977, August 2019.

[4] Leo Breiman. Statistical modeling: the two cultures (with comments and a rejoinder by the author). *Stat. Sci.*, 16(3):199–231, 2001.

[5] Katherine S Button, John P A Ioannidis, Claire Mokrysz, Brian A Nosek, Jonathan Flint, Emma S J Robinson, and Marcus R Munafò. Power failure: why small sample size undermines the reliability of neuroscience. *Nat. Rev. Neurosci.*, 14(5):365–376, May 2013.

[6] J Canny. A computational approach to edge detection. *IEEE Trans. Pattern Anal. Mach. Intell.*, 8(6):679–698, June 1986.

[7] Adriana Di Martino, David O'connor, Bosi Chen, Kaat Alaerts, Jeffrey S Anderson, Michal Assaf, Joshua H Balsters, Leslie Baxter, Anita Beggiato, Sylvie Bernaerts, and others. Enhancing studies of the connectome in autism using the Autism Brain Imaging Exchange II. *Sci. Data*, 4(1):1–15, 2017.

[8] Ron Eglash. Broken metaphor: the master-slave analogy in technical literature. *Technol. Cult.*, 48(2):360–369, 2007.

[9] Oscar Esteban, Daniel Birman, Marie Schaer, Oluwasanmi O Koyejo, Russell A Poldrack, and Krzysztof J Gorgolewski. MRIQC: advancing the automatic prediction of image quality in MRI from unseen sites. *PLoS One*, 12(9):e0184661, September 2017.

[10] Oscar Esteban, Christopher J Markiewicz, Ross W Blair, Craig A Moodie, A Ilkay Isik, Asier Erramuzpe, James D Kent, Mathias Goncalves, Elizabeth DuPre, Madeleine Snyder, Hiroyuki Oya, Satrajit S Ghosh, Jessey Wright, Joke Durnez, Russell A Poldrack, and Krzysztof J Gorgolewski. fMRIPrep: a robust preprocessing pipeline for functional MRI. *Nat. Methods*, 16(1):111–116, January 2019.

[11] Bruce Fischl. Freesurfer. *Neuroimage*, 62(2):774–781, 2012.

[12] Matthew F Glasser, Timothy S Coalson, Emma C Robinson, Carl D Hacker, John Harwell, Essa Yacoub, Kamil Ugurbil, Jesper Andersson, Christian F Beckmann, Mark Jenkinson, Stephen M Smith, and David C Van Essen. A multi-modal parcellation of human cerebral cortex. *Nature*, 536(7615):171–178, 08 2016.

[13] Krzysztof J Gorgolewski, Fidel Alfaro-Almagro, Tibor Auer, Pierre Bellec, Mihai Capotă, M Mallar Chakravarty, Nathan W Churchill, Alexander Li Cohen, R Cameron Craddock, Gabriel A Devenyi, Anders Eklund, Oscar Esteban, Guillaume Flandin, Satrajit S Ghosh, J Swaroop Guntupalli, Mark Jenkinson, Anisha Keshavan, Gregory Kiar, Franziskus Liem, Pradeep Reddy Raamana, David Raffelt, Christopher J Steele, Pierre-Olivier Quirion, Robert E Smith, Stephen C Strother, Gaël Varoquaux, Yida Wang, Tal Yarkoni, and Russell A Poldrack. BIDS apps: improving ease of use, accessibility, and

reproducibility of neuroimaging data analysis methods. *PLoS Comput. Biol.*, 13(3):e1005209, March 2017.

[14] Brian E Granger and Fernando Pérez. Jupyter: thinking and storytelling with code and data. *Comput. Sci. Eng.*, 23(2):7–14, March 2021.

[15] Harry F Harlow. The formation of learning sets. *Psychol. Rev.*, 56(1):51, 1949.

[16] Charles R Harris, K Jarrod Millman, Stéfan J van der Walt, Ralf Gommers, Pauli Virtanen, David Cournapeau, Eric Wieser, Julian Taylor, Sebastian Berg, Nathaniel J Smith, Robert Kern, Matti Picus, Stephan Hoyer, Marten H van Kerkwijk, Matthew Brett, Allan Haldane, Jaime Fernández del Río, Mark Wiebe, Pearu Peterson, Pierre Gérard-Marchant, Kevin Sheppard, Tyler Reddy, Warren Weckesser, Hameer Abbasi, Christoph Gohlke, and Travis E Oliphant. Array programming with NumPy. *Nature*, 585(7825):357–362, September 2020.

[17] Mahan Hosseini, Michael Powell, John Collins, Chloe Callahan-Flintoft, William Jones, Howard Bowman, and Brad Wyble. I tried a bunch of things: the dangers of unexpected overfitting in classification of brain data. *Neurosci. Biobehav. Rev.*, 119:456–467, December 2020.

[18] Brian Kernighan and Rob Pike. *The Unix Programming Environment*. Prentice Hall, 1984.

[19] Allison Koenecke, Andrew Nam, Emily Lake, Joe Nudell, Minnie Quartey, Zion Mengesha, Connor Toups, John R Rickford, Dan Jurafsky, and Sharad Goel. Racial disparities in automated speech recognition. *Proc. Natl. Acad. Sci. U. S. A.*, 117(14):7684–7689, April 2020.

[20] K S Krishnapriya, Vítor Albiero, Kushal Vangara, Michael C King, and Kevin W Bowyer. Issues related to face recognition accuracy varying based on race and skin tone. *IEEE Trans. Technol. Soc.*, 1(1):8–20, March 2020.

[21] Jingwei Li, Danilo Bzdok, Jianzhong Chen, Angela Tam, Leon Qi Rong Ooi, Avram J Holmes, Tian Ge, Kaustubh R Patil, Mbemba Jabbi, Simon B Eickhoff, B T Thomas Yeo, and Sarah Genon. Cross-ethnicity/race generalization failure of behavioral prediction from resting-state functional connectivity. *Sci. Adv.*, 8(11):eabj1812, March 2022.

[22] Robert C MacCallum, Shaobo Zhang, Kristopher J Preacher, and Derek D Rucker. On the practice of dichotomization of quantitative variables. *Psychol. Methods*, 7(1):19–40, March 2002.

[23] D Marr and E Hildreth. Theory of edge detection. *Proc. R. Soc. Lond. B Biol. Sci.*, 207(1167):187–217, February 1980.

[24] Thomas Naselaris, Emily Allen, and Kendrick Kay. Extensive sampling for complete models of individual brains. *Curr. Opin. Behav. Sci.*, 40:45–51, August 2021.

[25] Juan Nunez-Iglesias, Stéfan Van Der Walt, and Harriet Dashnow. *Elegant SciPy: The Art of Scientific Python*. O'Reilly Media, Inc., 2017.

[26] Nobuyuki Otsu. A threshold selection method from gray-level histograms. *IEEE Trans. Syst. Man Cybern.*, 9(1):62–66, 1979.

[27] Fernando Perez, Brian E Granger, and John D Hunter. Python: an ecosystem for scientific computing. *Comput. Sci. Eng.*, 13(2):13–21, 2010.

[28] Jonathan D Power, Kelly A Barnes, Abraham Z Snyder, Bradley L Schlaggar, and Steven E Petersen. Spurious but systematic correlations in functional connectivity MRI networks arise from subject motion. *Neuroimage*, 59(3):2142–2154, February 2012.

[29] Martin Reuter, M Dylan Tisdall, Abid Qureshi, Randy L Buckner, André J W van der Kouwe, and Bruce Fischl. Head motion during MRI acquisition reduces gray matter volume and thickness estimates. *Neuroimage*, 107:107–115, February 2015.

[30] Ariel Rokem and Kendrick Kay. Fractional ridge regression: a fast, interpretable reparameterization of ridge regression. *Gigascience*, November 2020.

[31] Galit Shmueli. To explain or to predict? *Stat. Sci.*, 25(3):289–310, 2010.

[32] Nitish Srivastava, Geoffrey Hinton, Alex Krizhevsky, Ilya Sutskever, and Ruslan Salakhutdinov. Dropout: a simple way to prevent neural networks from overfitting. *J. Mach. Learn. Res.*, 15(1):1929–1958, 2014.

[33] John D Van Horn, Jeffrey S Grethe, Peter Kostelec, Jeffrey B Woodward, Javed A Aslam, Daniela Rus, Daniel Rockmore, and Michael S Gazzaniga. The functional magnetic resonance imaging data center (fmridc): the challenges and rewards of large–scale databasing of neuroimaging studies. *Proc. R. Soc. Lond. B Biol. Sci.*, 356(1412):1323–1339, 2001.

[34] John Darrell Van Horn. Opinion: big data biomedicine offers big higher education opportunities. *Proc. Natl. Acad. Sci. U. S. A.*, 113(23):6322–6324, June 2016.

[35] Gaël Varoquaux. Cross-validation failure: small sample sizes lead to large error bars. *Neuroimage*, June 2017.

[36] Pauli Virtanen, Ralf Gommers, Travis E Oliphant, Matt Haberland, Tyler Reddy, David Cournapeau, Evgeni Burovski, Pearu Peterson, Warren Weckesser, Jonathan Bright, Stéfan J van der Walt, Matthew Brett, Joshua Wilson, K Jarrod Millman, Nikolay Mayorov, Andrew R J Nelson, Eric Jones, Robert Kern, Eric Larson, C J Carey, İlhan Polat, Yu Feng, Eric W Moore, Jake VanderPlas, Denis Laxalde, Josef Perktold, Robert Cimrman, Ian Henriksen, E A Quintero, Charles R Harris, Anne M Archibald, Antônio H Ribeiro, Fabian Pedregosa, Paul van Mulbregt, and SciPy 1.0 Contributors. SciPy 1.0: fundamental algorithms for scientific computing in python. *Nat. Methods*, 17(3):261–272, March 2020.

[37] Hadley Wickham. Tidy data. *J. Stat. Softw.*, 59(1):1–23, 2014.

[38] Tal Yarkoni and Jacob Westfall. Choosing prediction over explanation in psychology: lessons from machine learning. *Perspect. Psychol. Sci.*, 12(6):1100–1122, 2017.

[39] Jason D Yeatman, Brian A Wandell, and Aviv A Mezer. Lifespan maturation and degeneration of human brain white matter. *Nat. Comm.*, 5(1):1–12, 2014.

[40] Anastasia Yendiki, Kami Koldewyn, Sita Kakunoori, Nancy Kanwisher, and Bruce Fischl. Spurious group differences due to head motion in a diffusion MRI study. *Neuroimage*, 88:79–90, March 2014.

INDEX

Printed in the USA
CPSIA information can be obtained
at www.ICGtesting.com
JSHW050916211023
50486JS00001B/1